Books are to be returned on or before
the last date below.

Sir Henry Bessemer:
Father of the Steel Industry

Sir Henry Bessemer 1813–1898

Sir Henry Bessemer:
Father of the Steel Industry

edited by

C. BODSWORTH

Book 690
Published in 1998 by
IOM Communications Ltd
1 Carlton House Terrace
London SW1Y 5DB

IOM Communications Ltd
is a wholly-owned subsidiary of
The Institute of Materials

ISBN 1 86125 054 1

Printed and bound in the UK by
The University Press, Cambridge

Contents

Preface

This volume commemorates the centenary of the death of Sir Henry Bessemer, one of the outstanding inventors of the nineteenth century.

Sir Henry Bessemer was born in 1813 and died in 1898. He is mainly known as the inventor of the Bessemer converter, but this is only one of his many achievements. He was an outstanding inventor with 117 patents in his name, forty per cent of which were totally unrelated to the iron and steel industries.

His other inventions range from developing a die metal that could reproduce fine detail on cardboard and leather, a machine for extracting the juice from sugar cane to a machine for grinding and polishing lenses for telescopes.

The first chapter of this book deals in detail with Sir Henry Bessemer's life and career expanding upon the points mentioned above, and relating more interesting facts. The following five chapters are detailed descriptions of the use of the Bessemer process in various geographical areas of England and Wales. Each chapter explains the history of the Bessemer process in these areas. Included is the story of how the British Government dismissed his ideas for improved gunnery, resulting in Napoleon III of France financing his research. There is a chapter on the impact of the Bessemer process in the United States, including comments by Andrew Carnegie, the American entrepreneur, referring to Bessemer as the 'Great King of Steel'. The chapter on the development of the Bessemer process in Europe also talks about the Thomas process and the use of slag as fertiliser. The final chapter of the book 'A Bessemer Miscellany' is a collection of interesting information about Sir Henry Bessemer, as well as a look at how the Bessemer process has developed and what form it takes today. This includes memories of people who met Bessemer and were present at some of his experiments. The book is illustrated throughout with line drawings and photographs of steel works and other pictures of interest. Finally a comprehensive index is provided.

Sir Henry Bessemer, Inventor

C. BODSWORTH

The Bessemer converter process is recognised throughout the world as the first bulk steelmaking method. It made possible the production of homogeneous liquid metal with greater ductility than cast iron, with higher strength and toughness than wrought iron and in vastly greater quantities at much lower cost and in a fraction of the time required to make crucible steel. Commercial production of Bessemer steel began in 1858 and the output reached a peak in Great Britain in 1890. By 1920 only one works was still operating the process, but it continued as a major contributor to the total steel production in the USA, Germany, Belgium and France until the 1950s. By that time modifications of the original converter process were beginning to appear, using an oxygen blast in place of air and with all or part of the blast blown downwards onto the molten metal. This mode of operation had been proposed, but was not developed, by Bessemer. Modern variants of these developments now account for the bulk of the steel produced throughout the world. So the steel industry today is actually based on the derivatives of the original Bessemer process.

It is not generally recognised that the development of the converter process was only one of the many achievements of Sir Henry Bessemer. He was an outstanding inventor with 117 patents in his name, 40 per cent of which were totally unrelated to the iron and steel industries. Many of his inventions were not protected by patents. His autobiography[1] reveals a self-educated man with an immense capacity for work and a wide range of talents. There is ample evidence of his rapid recognition of the inefficient manufacturing methods which were in use at that time and of his self confidence that he could devise superior methods for achieving the same objectives.

He was a skilled designer and draughtsman, designing most of the equipment which was described in his patents and producing the detailed engineering drawings which were required for their construction. His inventive genius was combined with an astute and shrewd business sense. It is this unique combination of ingenuity, self motivation and business acumen which sets him apart from most of his contemporaries. He had a strong personality. He showed no diffidence in proclaiming his discoveries and displayed a belligerent reaction to any criticism of his claims, but this was counter balanced by a generous disposition and a dedication to his wife and three children.

The basis of his engineering and metal working skills was evidently acquired from his father, Anthony Bessemer, who was a highly skilled craftsman and a successful businessman. As a child, Anthony had been taken by his parents to live in Holland. After serving an engineering apprenticeship he moved to Paris, where he was employed by the Paris Mint. Whilst there, he designed a copying and engraving machine for making coining dies from artist's original patterns. This won international acclaim and he was rewarded with membership of the Academy of Sciences. The French revolution abruptly teminated his activities.

He and his wife managed to return to England, but most of their assets had to be left behind. After a few years in London regenerating his fortune, he bought a small country estate at Charlton, near Hitchen in Hertfordshire, where Henry Bessemer was born on the 19th January, 1813.

By the time Henry left school with only an elementary education at the age of 14, his father had built a type casting works in the grounds of his estate. With a sound financial income from this enterprise, there was no necessity for the son to find work to help to support the family. Henry therefore persuaded his father to let him work in the foundry and associated workshop, giving him scope for self education to develop his natural talents. With the aid of a small lathe and a ready supply of white metal, he was encouraged to make working models of moulding devices.

At the age of 17 the family business was moved to London, where he now began to make white metal reproductions of plaster casts. Experiments resulted in the development of modified plaster of Paris moulds which facilitated the reproduction as castings of very thin flower petals and leaves and which eliminated the porosity which was usually encountered with 'lost wax' castings. A technique was developed for copper plating the white metal castings to improve their appearance. The results were exhibited in 1832, several years before electrometallurgy was first described by Jordan and Spencer.

By experimentation, Bessemer developed a die metal which had a greater hardness than white metal, but which still had a moderately low melting point. Dies produced in this alloy could reproduce fine detail. Initially with a fly press and then with a hydraulic press, they were used to make sharp impressions on cardboard and on leather, which sold for a good profit. This was his first successful commercial venture. He also devised a method of making a die from an embossed pattern, which could then be used for multiple reproductions. The realisation that this type of die could be used to forge the Government embossed stamps, which were a charge on legal documents, led him to devise a die which could emboss the documents and simultaneously perforate the stamp to prevent reuse. The Government Stamp Office encouraged him in his efforts with the prospect of employment. At the suggestion of his fiancee, however, he devised a more simple method of incorporating a date stamp within the existing dies. This method was adopted as standard practice by the Stamp Office, but no payment was made to Bessemer for his efforts and the theft of his idea remained a source of contention with him for the next forty years.

He was married in 1833 and set up home initially in Northampton Square but soon moved to Baxter House, St. Pancras, where the grounds gave him more room to develop his experimental and production workshops. The family moved at a later date to Highgate and finally to Denmark Hill, all the homes being in London.

Bessemer continued to produce innovative developments in casting practice. A procedure was devised to avoid the rejection of printing type due the formation of porosity during solidification. This comprised the use of water cooled moulds and the application of a vacuum to the mould during injection moulding. Anticipating modern methods of semi-solid processing, he formulated a new hard white metal alloy composition which had a wide freezing range, allowing the production of fine detail on stamping dies and embossing rolls by the application of pressure during the final stages of solidification.

Meanwhile, Bessemer had started to design a wide range of devices which were intended to reduce the costs of manufacture of a variety of artefacts. The first of these was a mechanical saw for cutting the thin slices of plumbago required for the production of pencil leads, replacing a laborious hand cutting operation. Recognising that a large proportion of the valuable mineral was lost as sawdust and broken fragments, a method was devised for consolidating the powder with a binder under pressure, which became standard practice for the industry. A mechanical composing machine was designed and constructed, which was capable of setting up print type at more than three times the rate achieved by hand composing. Although the machine was used for a period in a newspaper office, it was too far ahead of the time and was soon scrapped.

A more profitable activity was the development of a method for the production of imitation figured Genoa velvet by embossing a permanent pattern on plain velvet. This was eventually achieved by passing the velvet between hollow rolls on which the pattern was engraved. A permanent pattern was only obtained if the rolls were warmed, so they were heated internally by a gas burner. The actual roll temperature was critical and had to be controlled to within ± 10 °C. Since optical pyrometers and contact thermocouples had not been invented, Bessemer prepared three fusible alloys with melting points spanning the working temperature range. These were pressed against the rolls to indicate whether more or less heat was required and acted in similar manner to the 'Sentinel' cones which Harry Brearley introduced about 80 years later.

Up to this stage, Bessemer's numerous innovations had produced only a moderate income. The callous treatment which he had received from the Stamp Office for his date stamp and the limited returns from several of his other inventions made him aware of the need to seek patent protection before revealing the details of his discoveries. His first patent, dated 8th March 1838, related to the casting, breaking off and counting of printing type. Three years elapsed, however, before the next patent was submitted, during which time he was fully occupied with his first really successful financial venture.

He was surprised to discover the very high cost of the 'gold' powder which was sold for decorative lettering and japanning. Tests quickly showed that it was a simple brass or bronze alloy, which was not an expensive material in bulk quantities but sold for about £11 per kilogram in fine powder form. Enquiries revealed that the powder was produced entirely by laborious hand methods, so Bessemer conducted extensive investigations to devise machines which would produce a similar powder. When this had been accomplished, he found that the estimated cost of production was only about one twentieth of the traditional method. He also realised that a patent would provide no protection, since it would be difficult to prove that a supply of powder had been produced by his method. Accordingly, he set up a manufacturing unit in the grounds of his Baxter House. The engineering drawings of the machine parts were prepared by Bessemer and sent to various manufacturers for construction in such a way that no supplier could understand the function of any of the machines. His brothers-in-law were recruited as operators and were the only persons, apart from Bessemer, who were allowed to enter the premises (the boiler man had no access to the production plant) so all the machines had to be made self activating as far as possible.

An ingeneous method of air separation was devised for dividing the powder product into different size fractions. The colour of the powder was varied by closely controlled surface oxidation in a heated drum. The range of colours was extended by alloying pure copper with a variety of elements, including nickel, molybdenum and tungsten. A white finish was imparted by the deposition from aqueous solution of a thin layer of tin. The plant started in full production in 1843, shortly after his 30th birthday. It produced a good income which financed his innovations over the next two decades. The procedure remained a closely guarded secret for about 35 years and he eventually assigned the process to one of his brothers in law, Richard Allen. Hitherto, the metal flakes had only been used in dry powder form. By trial and error, Bessemer discovered a suitable solution in which the powder could be mixed without discolouration and applied as a paint. This was the only aspect of his powder work which was covered by a patent, granted a few days before his 31st birthday.

By this time an office had been opened in London. Mr Robert Longsdon was employed as a draughtsman to assist with the rapidly increasing volume of work, but he soon became a friend and eventually a partner in Bessemer's activities.

The patents which he acquired during the next five years amply demonstrate the amazing breadth of his interests. One of these was to provide 'elastic communication between railway carriages so that a train is continuous from end to end'. This forerunner of the modern corridor train had collapsible hoods fitted at the ends of all the carriages, permitting passenger transfer from coach to coach when the hoods were joined together. Another patent covered a steam powered fan and centrifugal pump for the provision of ventilation and the drainage of water from mines.

An interest in the possibility of developing a solar furnace (a topic to which he returned in retirement) led to a series of patents on the production of glass. The initial work was concerned with the manufacture of optical glass for the lenses which were required for the furnace. Only relatively small diameter disks of homogeneous glass were available commercially. Cold model experiments with a viscous transparent fluid revealed the difficulty of obtaining homogeneity in a highly viscous mass. Stirring introduced dissolved air which escaped only slowly when the agitation ceased and it had not entirely escaped before density differences reimposed a composition gradient, which affected the optical properties. A novel system which involved the slow rotation of a heated conical crucible was found to produce glass which was sufficiently homogeneous for the production of large lenses.

During the course of these experiments, Bessemer noted the relatively slow rate of fusion of the coarse powders which were used commercially for glass manufacture. Reasoning that the rate of reaction of the silica with the lime and soda was dependent on the surface area, he demonstrated that fine grinding and thorough mixing of the constituents accelerated the speed of both fusion and homogenisation to form a fluid glass.

All the glass which was made at this time was produced by fusion in crucibles. For his first foray into the development of methods for bulk production at elevated temperatures, Bessemer designed a reverberatory furnace for the mass production of sheet glass. This was built in 1846. A major problem was the contamination of the glass by drops of molten refractory, which were fluxed from the furnace roof by alkali vapours in the atmosphere above the bath of molten glass. A solution was found by using a hollow box section of refractory brick for the construction of the roof arch. The glass was formed in one third of the time and with only one third of the fuel consumption required in crucible melting. A slot along one side of the melting tank was closed with an iron bar during the fusion period. The bar was removed when the molten glass was ready for discharge, allowing the glass to flow over a lip and down an incline between two rollers where it was compressed to the required thickness and sheared into suitable lengths. The sheets were collected on a flat bed at the foot of the incline. The patent rights for the glass furnace were sold to the Chance Glass Company for £6000. Later plans to open his own glass works to manufacture plate glass using a circular reverberatory furnace were abandoned when he was unable to find suitable partners to join in the venture.

A vacuum table to hold securely a sheet of plate glass during surface grinding and polishing operations was demonstrated at the International Exhibition of 1851, together with a number of Bessemer's other innovations. Although the vacuum table vastly simplified the working procedure, it was one of several Bessemer patents which was not taken up by industry. In similar vein, a method which he devised for the production of glass mirrors using a silver amalgam was overtaken by the introduction of a more simple aqueous process before the amalgam method had been widely adopted commercially.

In 1849 the Society of Arts announced a competition with the award of a gold medal to the person within the next twelve months who achieved the greatest improvement in the amount of juice extracted from raw sugar cane. Bessemer counted a Jamaican sugar planter amongst his acquaintances and was astonished to learn about the crude methods which were used to extract the juice. He decided to enter the competition. With no detailed knowledge of the problem, he rejected the conventional roller mill and designed a reciprocating steam engine which sheared short lengths from the sugar cane on each stroke and compressed the cut pieces repetitively in a perforated chamber. A centrifugal filter was incorporated to clarify the liquor. The yield of juice was increased markedly yet, in contrast to the existing equipment, the new machine was much lighter and could be transported easily round the plantations. The gold medal was presented to him by Prince Albert. Thirteen patents on sugar extraction and refining were issued to Bessemer during the next four years and some of these devices were included in the 1851 exhibition.

The railway system was now expanding rapidly in Great Britain and periodically Bessemer was attracted to the developments, as witnessed again by his patents. In December 1853 he patented a design for the first system of hydraulic brakes which could be applied to all the wheels on a train, but it was too advanced for the time and was not adopted by the railways. Other patents were concerned with the manufacture of railway wheels and rails. To an increasing extent, therefore, Bessemer was using ferrous materials for the artefacts which he was designing and he must have been well aware of the strengths and limitations of the iron and steel then available. The limitations in particular frustrated the achievement of one of his new interests and resulted in the development of his steelmaking process.

This interest arose at the start of the Crimean war, with the recognition of the low accuracy achieved by the ordnance then in use. Round shot was starting to be replaced by elongated shells, but they were all fired from smooth bore guns. Realising that the effectiveness would be increased if the shell could be made to rotate about its longitudinal axis, Bessemer designed a shell containing a longitudinal channel which terminated in a tangential vent. He reasoned that the escape of a small portion of the propulsion gases through the vent would impart a rotary motion to the shell. The British War Department was still using only round shot and showed no interest in the development of better ammunition. So Bessemer made a small cast iron mortar with which he proved the validity of his concept by firing shells within the grounds of his Baxter House home. The War Department was not impressed, but a fortuitous meeting with Prince Napoleon (later Napoleon the Third of France) resulted in sponsorship of the development work by the Prince, who opened an account with Coutts Bank from which Bessemer could draw funds to finance his experiments. Full scale tests proved the increased accuracy and penetrating power of the shells, but Bessemer found that the cast iron cannon then in use were too weak to withstand the pressures arising during the discharge of the heavier projectiles. Although he admitted to 'very little knowledge of iron metallurgy' Bessemer resolved to produce a superior form of ferrous metal which would be suitable for the manufacture of stronger guns. The converter process was his solution of this objective, but a writer of his obituary notice remarked that 'he would not have invented the Bessemer process if he had been taught metallurgy'. The field tests on the shells were completed in late December 1854 and the need for stronger guns was then confirmed. Bessemer's first patent for 'improvements in the manufacture of iron and steel' was dated only three weeks later.

The initial patent described the refining of pig iron in a reverberatory furnace, similar in design to the one which he had used earlier for the production of sheet glass but with one major modification. Air was blown through perforations in the fire bridge to increase the temperature by burning the remaining combustible gases above the metal bath. Perseverance with this concept might have pre-empted Siemens development of the open hearth furnace. A chance observation, however, that two seemingly unmelted blocks of pig iron on the side of the hearth were merely shells of decarburised iron led to the realisation that air could be used both to oxidise the impurities from molten pig iron and to raise the metal temperature without any extraneous fuel supply.

In the summer of 1856 he made a public announcement of his new process in a paper entitled 'The manufacture of malleable iron and steel without fuel' which he presented at the Cheltenham meeting of the British Association (see appendix). Following the disastrous early attempts to produce satisfactory metal, he continued to experiment and, in 1859, read a paper 'On the manufacture of malleable iron and steel' at a meeting of the Institute of Civil Engineers. This described how sound metal could be produced when a pig iron made from hematite ore was used and advocated the use of the Bessemer steel for numerous purposes, including guns, ship plate, rails and bridges. A wide variety of applications was demonstrated at the 1862 International Exhibition, including the first steel nails and steel wire.

A stationary upright vessel, rather like a miniature cupola, with a ring of tuyeres set in the lower wall was adopted as the first version of the Bessemer converter (Fig. 1). Several modifications of the design of the fixed converter were described before the more familiar tilting, bottom blown converter was introduced in a patent dated 1st March 1860, shortly after Bessemer opened his steelworks in Sheffield (Fig. 2). Over thirty patents protected the developments in the process, the equipment and the production of ingots. The casting pit employed bottom-teaming ladies which were carried on a rotary arm from the converter vessels to the ingot moulds. The latter were provided with hydraulic rams for ejection of the solidified ingots.

A patent was obtained in 1846 for the continuous casting of tin and lead sheet, using an apparatus very similar to that which was described in the same year for the production of sheet glass. Equipment for the continuous casting of steel strip was described in an 1856

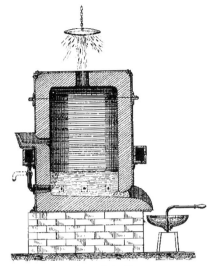

Fig. 1 First version of the Bessemer converter.

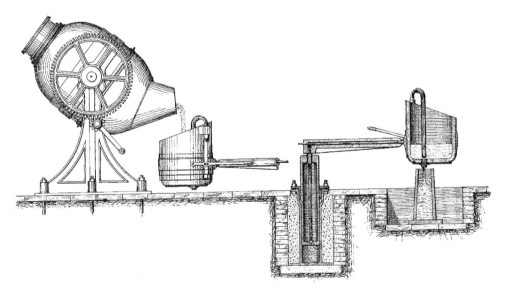

Fig. 2 Bottom blown Bessemer converter.

patent. The molten metal was poured into a narrow gap between two slowly rotating water cooled rolls. The emerging solid metal passed between driven guide rolls and cut-off shears on a curved track to discharge onto a horizontal bed, in similar manner to the continuous lead and glass production equipment (Fig. 3). The apparatus does not appear to have progressed beyond the pilot plant stage, mainly because it proved impossible to maintain a uniform width of solidified metal emerging from the rolls. Thirty five years later, in one of only two

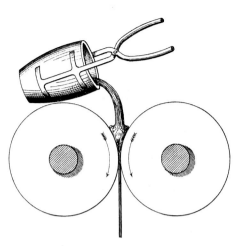

Fig. 3   Bessemer's initial continuous casting equipment using chilled-rolls.

papers which he presented to the Iron and Steel Institute,[2] he described how a tundish with a series of holes in the base could be mounted above the roll gap to provide a uniform feed and produce a constant strip width (Fig. 4). Almost 100 years elapsed from the date of the patent to the commercial application of continuous casting to steel and even longer before the water cooled roll found application in the production of rapidly solidified ribbons. The twin roll caster is now the principle method which is undergoing development internationally for the continuous production of thin steel sheet.

When the Bessemer Sheffield works had demonstrated the capabilities of the converter and steelmakers around the world were taking out licences to operate the process, Bessemer relaxed his inventive activities. From a peak of 14 patents in 1855 and 10 in 1856, his output declined to an average of only three or four per annum up to his retirement in 1873. A number of these related to the manufacture of guns and projectiles. His autobiography contains lengthy diatribes concerning the refusal by the British War Department to consider the use of Bessemer metal for these purposes, although a considerable number of guns were produced by Bessemer for overseas markets. He apparently regarded the cold ductility of the metal as its principle asset, with strength of secondary importance. The significance of impact resistance to withstand explosive detonations was not recognised at that time.

The other patents which were taken out during this period demonstrated the continuing breadth of his interests. In addition to steel production, they included the construction of hydraulic presses and blowing engines for blast furnaces, the manufacture of grindstones, firebricks, retorts and crucibles, the design of buildings for steelworks and the construction of asphalt pavements.

One group of patents related to a concept which absorbed a considerable sum of money, but was not carried through to completion. It appears that Bessemer suffered acutely from sea sickness and he was severely incapacitated after a channel crossing in 1868. This led him to design a steam ship saloon which was to be mounted mid ship and free to move on an axis parallel to the keel. A steersman would control the extent of the cabin movement as the ship

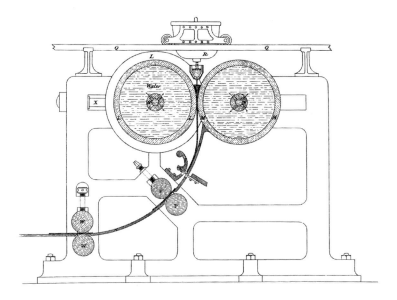

Fig. 4 Bessemer's patented apparatus for continuous casting using water cooled rolls

rolled by operation of hydraulic dampers. An extended length of the hull and a low freeboard would minimise the pitching motion. The Bessemer Saloon Ship Company was formed to build a cross channel ferry incorporating these designs, but the Company ran into financial difficulties before the construction was completed. Two trial trips were made before the hydraulic dampers had been installed and the cabin was not free to move. The ship proved difficult to manoeuvre and on each occasion it collided with a harbour pier, causing damage which was so extensive that the Company was forced into liquidation by the cost of the repairs. The suspended cabin was never given a proper test.

Bessemer retired at the age of 60, but he did not rest his talents. He made extensive alterations to his house and gardens in Denmark Hill (Figs 5 and 6) and started to build an elaborate observatory and reflecting telescope in the grounds (Fig. 7). A machine was made to grind and polish the lenses for the telescope, but continuing improvements to the design failed to produce the accuracy of the lens profile required for astronomical purposes. Construction of the telescope became an absorbing passtime which was never completed. A further attempt to build a solar furnace for melting the more refractory metals was also frustrated by his inability to grind accurate profiles on the solar reflectors with the equipment available at that time (Figs 8 and 9). He succeeded only in melting copper and vapourising zinc. One of his last activities was to finance a diamond cutting works for one of his grandsons. The cutting and polishing machines were all designed by Bessemer.

After a long and active retirement, Bessemer died on the 15th March 1898. His achievements and inventions had been rewarded by international recognition and honours had been showered upon him. He was appointed an honorary member of the Iron Board for Sweden, a Knight Commander of the Order of His Imperial Majesty Franz Josef (of Austria) and a

Fig. 5   Bessemer's house Denmark Hill.

Fig. 6   An internal picture of Bessemer's house Denmark Hill.

Fig. 7   Bessemer's observatory.

Freeman of the City of Hanover. The Emperor Napoleon the Third awarded him the Grand Cross of the Legion of Honour. The British Government initially refused permission for him to accept the award. This provoked an acrimonious attack on the Government by Bessemer in the Times newspaper, in which the failure to recompense him for his method of preventing forgery of Government stamps was expounded at great length, The issue was resolved by the award of a Knighthood, which was conferred on him in 1879. He was elected a member of the Society for the Encouragement of the National Industry of Paris and of the Royal Academy of Trade in Berlin.

Nearer to home he was awarded the Freedoms of the City of London, the Cutlers Company of London and the Turners Company. He was elected a Fellow of the Royal Society and a member of the Royal Society of British Architects and the Society of Mechanical Engineers of England and America. One of the first awards which he received was the Telford Gold Medal of the Institute of Civil Engineers for the paper which he presented at their meeting in 1859. At a later date he was awarded the Howard prize and became a member of the council of the ICE.

The Iron and Steel Institute was formed in 1869 and Bessemer was one of the founder members. He succeeded the Duke of Devonshire as the second President of the Institute and held office in 1871–1873. During this tenure, he provided a sum of money to endow the

Fig. 8 Bessemer's solar furnace.

annual award of the Bessemer Gold Medal for outstanding contributions to the Iron and Steel industry. The medal became the premier award of the Iron and Steel Institute and its successors, the Metals Society, the Institute of Metals and now the Institute of Materials. The long list of international distinguished metallurgists who have been pleased to accept the award of the medal is continuing proof of the esteem in which this pioneer of bulk steel manufacture is still held within the profession.

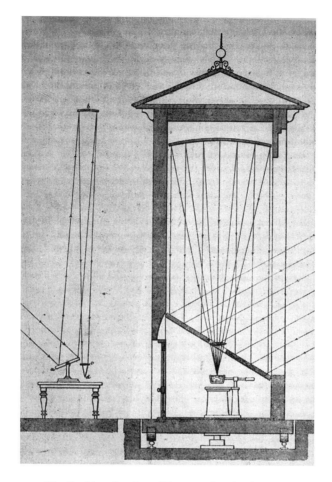

Fig. 9   Line drawing of Bessemer's solar furnace.

REFERENCES

1.   *Sir Henry Bessemer, FRS: An autobiography*, 1989, The Institute of Metals.
2.   Henry Bessemer: *Journal Iron Steel Inst.*, 1891, 23–41.

Appendix

PAPER READ BY BESSEMER BEFORE SECTION G OF THE BRITISH ASSOCIATION AT CHELTENHAM IN 1856

'For the last two years, my attention has been exclusively directed to the manufacture of malleable iron and steel, in which, however, I had made but little progress until within the last eight or nine months. The constant pulling down and rebuilding of furnaces, and the toil of daily experiments with large charges of iron, had already begun to exhaust my stock of patience; but the numerous observations I had made during this unpromising period, all tended to confirm an entirely new view of the subject which at that time forced itself upon my attentien, viz. that I could produce a much more intense heat, without any furnace or fuel, than could be obtained by either of the modifications I had used; and consequently, that I should not only avoid the injurious action of mineral fuel on the iron under operation, but that I should at the same time avoid also the expense of the fuel.

Some preliminary trials were made on from 10 lbs. to 20 lbs. of iron, and although the process was fraught with considerable difficulty, it exhibited such unmistakable signs of success as to induce me at once to put up an apparatus capable of converting about 7 cwt. of crude pig iron into malleable iron in thirty minutes. With such masses of metal to operate on, the difficulties which beset the small laboratory experiments of 10 lbs. entirely disappeared. On this new field of inquiry I set out with the assumption that crude iron contains about 5 per cent. of carbon; that carbon cannot exist at a white heat in the presence of oxygen, without uniting therewith and producing combustion; that such combustion would proceed with a rapidity dependent on the amount of surface of carbon exposed; and lastly, that the temperature which the metal would acquire would be also dependent on the rapidity with which the oxygen and carbon were made to combine; and consequently, that it was only neccssary to bring the oxygen and carbon together, in such a manner that a vast surface should be exposed to their mutual action, in order to produce a temperature hitherto unattainable in our largest furnaces.

With a view of testing practically this theory, I constructed a cylindrical vessel of 3 feet in diameter and 5 feet in height, somewhat like an ordinary cupola furnace. The interior is lined with firebricks, and at about 2 inches from tbe bottom of it I inserted five tuyere pipes, the nozzles of which are formed of well-burned fireclay, the orifice of each tuyere being about three-eighths of an inch in diameter; they are so put into the brick lining (from the outer side) as to admit of their removal and renewal in a few minutes when they are worn out. At one side of the vessel, about half-way up from the bottom, there is a hole made for running-in the crude metal, and on the opposite side there is a tap-hole, stopped with loam, by means of which the iron is run out at the end of the process. In practice, this converting vessel may be made of any convenient size, but I prefer that it should not hold less than one or more than five tons of fluid iron at each charge; the vessel should be placed so near to the discharge hole of the blast furnace as to allow the iron to flow along a gutter into it. A small blast cylinder is required capable of compressing air to about 8 lbs. or 10 lbs. to thc square inch. A communication having been made between it and the tuyeres before named, the converting vessel will be in a condition to commence work; it will, however, on the occasion of its first being used, after re-lining with firebricks, be necessary to make a fire in the interior with a few baskets of coke, so as to dry the brickwork and heat up the vessel for thc first operation, after which the fire is to be all carefully raked out at the tapping-hole, which is again to he made good with loam: the vessel will then be in readiness to commence work, and may be so continued, without any use of fuel, until the brick lining, in the course of time, becomes worn away, and a new lining is required. I have before mentioned that the tuyeres are situated nearly close to the bottom of the vessel, the fluid metal will therefore rise some 18 inches or 2 feet above them; it is therefore necessary, in order to prevent the metal from entering the tuyere holes, to turn on the blast before allowing the fluid crude iron to run into the vessel from the blast furnace. This having been done, and the metal run in, a rapid boiling up of the metal will be heard going on within the vessel, the metal being tossed violently about and dashed from side to side, shaking the vessel by the force with which it moves; from the throat of the converting vessel flame will immediately issue, accompanied by a few bright sparks, such as are always seen rising from the metal while running in to the pig-beds. This state of things will continue for about fifteen minutes, during which time the oxygen in the atmospheric air combines with the carbon contained in the iron, producing carbonic oxide, or carbonic acid gas, and at the same time evolving a powerful heat. Now, as this heat is generated in the interior of, and is diffused in innumerable fiery bubbles through the whole fluid mass, the metal absorbs the greater part of it, and its temperature becomes immensely increased; and by the expiration of the fifteen minutes before named, that part of the carbon which appears mechanically mixed and diffused throughout the crude iron has been entirely consumed; the temperature, however, is so high that the chemically combined carbon now begins to separate from the metal, as is at once indicated by an immense increase in the volume of flame rushing out of the throat of the vessel. The metal in the vessel now rises several inches above its natural level, and a light frothy slag makes its appearance, and is thrown out in large foam-like masses. This violent eruption of cinder generally lasts about five or six minutes, when all further appearance of it ceases, a steady and powerful flame replacing the shower of sparks and cinder which always accompanies the boil. The rapid union of carbon and oxygen which thus takes place adds still further to the temperature of the metal, while the diminished quantity of carbon present allows a part of the oxygen to combine with the iron, which undergoes combustion, and is converted into an oxide. At the excessive temperature that the metal has now acquired, the oxide, as soon as formed, undergoes fusion, and forms a powerful solvent of those earthy bases that are associated with the iron; the violent ebullition which is going on mixes most intimately the scoriae and metal, every part of which is thus brought in contact with the fluid oxide, which will thus wash and cleanse the metal most thoroughly from the silica and other earthy bases which are combined with the crude iron, while the sulphur and other volatile matters which cling so tenaciously to iron at ordinary temperatures are driven off, the sulphur combining with the oxygen, and forming sulphuric acid gas.

The loss in weight of crude iron during its conversion into an ingot of malleable iron was found, on a mean of four experiments, to be 12.5 per cent., to which will have to be added the loss of metal in the finishing rolls. This will make the entire loss probably not less than 18 per cent., instead of about 28 per cent. which is the loss on the present system. A large portion of this metal is, however, recoverable, by treating with carbonaceous gases the rich oxides thrown out of the furnace during the boil. These slags are found to contain innumerable small grains of metallic iron, which are mechanically held in suspension in the slags, and may be easily recovered.

I have mentioned that, after the boil has taken place, a steady and powerful flame succeeds, which continues without any change for about ten or twelve minutes, when it rapidly falls off. As soon as this diminution of flame is apparent, the workman will know that the process is completed, and that the crude iron has been converted into pure malleable iron, which he will form into ingots of any suitable size and shape by simply opening the tap-hole of the converting vessel, and allowing the fluid malleable iron to flow into the iron ingot moulds placed there to receive it. The masses of iron thus formed will be perfectly free from any admixture of cinder, oxide, or other extraneous matters, and will be far more pure, and in a forwarder state of manufacture, than a pile formed of ordinary puddle-bars. And thus it will be seen, that by a single process, requiring no manipulation or particular skill, and with only one workman, from three to five tons of crude iron pass into the condition of several piles of malleable iron in from thirty to thirty-five minutes, with the expenditure of about a third part the blast now used in a finery furnace, with an equal charge of iron, and with the consumption of no other fuel than is contained in the crude iron.

To those who are best acquainted with the nature of fluid iron, it may be a matter of surprise that a blast of cold air forced into melted crude iron is capable of raising its temperature to such a degree as to retain it in a perfect state of fluidity after it has lost all its carbon, and is in the condition of malleable iron, which, in the highest heat of our forges, only becomes softened into a pasty mass. But such is the excessive temperature that I am able to arrive at with a properly shaped converting vessel and a judicious distribution of the blast, that I am enabled not only to retain the fluidity of the metal, but to create so much surplus heat as to remelt all the crop-ends, ingot-runners, and other scrap that is made throughout the process, and thus bring them, without labour or fuel, into ingots of a quality equal to the rest of the charge of new metal. For this purpose a small arched chamber is formed immediately over the throat of the converting vessel, somewhat like the tunnel head of the blast furnace. This chamber has two or more openings on the sides of it, and its floor is made to slope downwards to the throat, as soon as a charge of fluid malleable iron has been drawn off from the converting vessel. The workman will take the scrap intended to be worked into the next charge, and proceed to introduce the several pieces into the small chamber, piling them up around the opening of the throat.

When this is done, he will run in his charge of crude metal, and again commence the process. By the time the boil commences, the bar-ends, or other scrap, will have acquired a white heat, and by the time it is over, most of them will have been melted and run down into the charge; any pieces, however, that remain, may then be pushed in by the workman, and by the time the process is completed, they will all be melted and intimately combined with the rest of the charge, so that all scrap iron, whether cast or malleable, may thus be used up without any loss or expense. As an example of the power that iron has of generating heat in this process, I may mention a circumstance that occurred to me during my experiments: I was trying how small a set of tuyeres could be used, but the size chosen proved to be too small, and after blowing into the metal for an hour and three-quarters, I could not get up heat enough with them to bring on the boil. The experiment was therefore discontinued, during which time two-thirds of the metal solidified, and the rest was run off. A larger set of tuyere pipes was then put in, and a fresh charge of fluid iron run into the vessel, which had the effect of entirely remelting the former charge, and when the whole was tapped out it exhibited, as usual, that intense and dazzling brightness peculiar to the electric light.

I have before mentioned that the fluid malleable iron is run into iron ingot moulds, leaving it to be inferred that ordinary ingot moulds may be used, or such as are generally employed by cast-steel makers; but a little consideration will show that such moulds would entail an amount of labour that is most desirable to avoid in the manufacture of iron. It is also most essential to remove the ingot to the rolls while still retaining a very high temperature, and thus avoid the re-heating which would otherwise be required. For this purpose, the moulds are placed on end in a vertical position, their insides being planed truly parallel; the bottom of the mould is movable,and is attached to a small plunger, like the ram of an ordinary hydraulic press, the cylinder of which is attached to the mould; a force-pump is worked by the nearest steam-engine, and has a pipe which leads from it to the cyclinder, so that when the mould is filled with metal, and the central part is still almost fluid, a cock is opened by the workman, which allows the water from the force-pump to raise the ram and force out the ingots, while still at a glowing white heat. The moulds are sunk below the surface of the ground, and a rail-track extends on each side of them, on which there is an iron truck, the under side of which is so formed as to receive the ingots as they are pushed out by the ram, which is then lowered by reversing the cock, and allowing the water to escape. The truck is then quickly rolled along the line of rails, taking with it the ingots to the rolls, the whole operation requiring only from one to two minutes, in which time the ingots will not have cooled down sufficiently in the centre for rolling, so that the first bars will be produced and finished off fit for sale wholly without the use of fuel, and within a period of forty minutes from the time of tapping the crude iron from the blast furnaces. If the iron is made in very large quantities, the heat of some of the ingots will not be retained until they can be rolled into bars; a small oven, capable of retaining the heat, must, in that case, be erected near the rolls, in which the ingots may remain until the rolls are at liberty. As soon as the workman has recharged the converting vessel, he will open another communication with the force-pump before mentioned, by means of which the moulds will be filled with water, and their temperature reduced; after which the water is to be run off, care being taken that the moulds are not, however, made so cold as to prevent them from drying off before the next charge of metal is ready. In this manner, the process of converting the crude into malleable iron, and the formation of it into ingots, may be carried on continuously throughout the day, at intervals of about three-quarters of an hour.

To persons conversant with the manufacture of iron, it will be at once apparent that the ingots of malleable metal which I have described will have no hard or steely parts, such as is found in puddled iron, requiring a great amount of rolling to blend them with the general mass, nor will such ingots require an excess of rolling to expel cinder from the interior of the mass, since none can exist in the ingot, which is pure and perfectly homogeneous throughout, and hence requires only as much rolling as is necessary for the development of fibre; it therefore follows that, instead of forming a merchant bar, or rail, by the union of a number of seperate pieces welded together, it will be far more simple and less expensive to make several bars or rails from a single ingot. Doubtless this would have been done long ago, had not the whole process been limited by the size of the ball which the puddler could make.

The facility which the new process affords of making large masses, will enable the manufacturer to produce bars that, in the old mode of working, it was impossible to obtain; while, at the same time, it admits of the use of more powerful machinery, whereby a great deal of labour will be saved, and the process be greatly expedited. I merely mention this fact in passing, as it is not my intention at the present moment to enter upon any details of the improvements I have made in this department of the manufacture, because the patents which I have obtained for them are not yet specified. Before, however, dismissing this branch of the subject, I wish to call the attention of the meeting to some of the peculiarities which distinguish cast steel from all other forms of iron – viz. the perfectly homogeneous character of the metal, the entire absence of sand-cracks or flaws, and its greater cohesive force and elasticity, as compared with the blister steel from which it is made; qualities which it derives *solely* from its fusion and formation into ingots, all of which properties malleable iron acquires, in like manner, by its fusion and formation into ingots in the new process; nor must it be forgotten that no amount of rolling will give to blister steel (although formed of rolled bars) the same homogeneous character that the cast steel acquires, by a mere extension of the ingot to some ten or twelve times its original length.

One of the most important facts connected with the new system of manufacturing malleable iron is, that all the iron so produced will be of that quality known as charcoal iron – not that any charcoal is used in its manufacture, but because the whole of the processes following the melting of it are conducted entirely without contact with, or the use of, any mineral fuel. The iron resulting therefrom will in consequence be entirely free from those injurious properties which that description of fuel never fails to impart to iron brought under its influence. At the same time, this system of manufacturing malleable iron offers extraordinary facilities for making large shafts, cranks, and other heavy masses. It will be obvious that any weight of metal that can be found in ordinary cast iron by the means at present at our disposal, may also be found in molten malleable iron, and be wrought into the forms and shapes required, provided that we increase the size and power of our machinery to the extent necessary to deal with such large masses of metal.

The manufacturer has gone on increasing the size of his smelting furnaces, and adopting to their use blast apparatus of the requisite proportions, and he has by this means lessened the cost of production in every way. His large furnaces require a great deal less labour to produce a given weight of iron than would have been required formerly to produce it with a dozen furnaces; and in like manner he diminishes his cost of fuel, blast, and repairs, while he ensures a uniformity in the result that never could have been arrived at by the use of a multiplicity of small furnaces. While the manufacturer has thus shown himself fully alive to these advantages, he has still been under the necessity of leaving the succeeding operations to be carried out on a scale wholly at variance with the principle he has found so advantageous in the melting department. It is true that, hitherto, no better method was known than the puddling process, in which from 4 to 5 cwt. of iron is all that can be operated upon at a time, and even this small quantity is divided into homoeopathic doses of some 70 lbs. or

80lbs., each of which is moulded and fashioned by human labour, and carefully watched and tended in the furnace, and removed therefrom one at a time, to be again carefully manipulated and squeezed into form. When we consider the vast extent of the manufacture, and the gigantic scale on which the early stages of the process are conducted, it is astonishing that no effort should have been made to raise the after processes somewhat nearer to a level commensurate with the preceding ones, and thus rescue the trade from the trammels which have so long surrounded it.

Before concluding these remarks, I beg to call your attention to an important fact connected with the new process, which affords peculiar facilities for the manufacture of cast steel.

At the stage of the process immediately following the boil, the whole of the crude iron has passed into the condition of cast steel of ordinary quality. By the continuation of the process, the steel so produced gradually loses its small remaining portion of carbon, and passes successively from hard to soft steel, and from soft steel to steely iron, and eventually to very soft iron; hence, at a certain period of the process, any quality of metal may be obtained. There is one in particular which, by way of distinction, I call semi-steel, being in hardness about midway between ordinary cast steel and soft malleable iron. This metal possesses the advantage of much greater tensile strength than soft iron; it is also more elastic, and does not readily take a permanent set, while it is much harder, and is not worn or indented so easily as soft iron; at the same time it is not so brittle or hard to work as ordinary cast steel. These qualities render it eminently well adapted to purposes where lightness and strength are specially required, or where there is much wear, as in the case of railway bars, which, from their softness and lamellar texture, soon become destroyed. The cost of semi-steel will be a fraction less than iron, because the loss of metal that takes place by oxidation in the converting vessel is about 2.5 per cent. less than it is with iron; but as it is a little more difficult to roll, its cost per ton may fairly be considered to be the same as iron; but as its tensile strength is some 30 or 40 per cent. greater than bar iron, it follows that, for most purposes, a much less weight of metal may be used to that taken in that way. The semi-steel will form a much cheaper metal than any that we are at present acquainted with.

In conclusion, allow me to observe that the facts which I have had the honour of bringing before the meeting have not been elicited fron mere laboratory experiments, but have been the result of working on a scale nearly twice as great as is pursued in our largest ironworks, the experimental apparatus doing 7cwt. in thirty minutes, while the ordinary puddling furnace makes only 4.5 cwt. in two hours, which is made into six separate balls, while the ingots or blooms are smooth, even prisms, 10 inches square by 30 inches in length, weighting about equal to ten ordinary puddle balls. A small portion of one of these ingots will be observed among the samples present; there is also a cylindrical mass of highly crystallised iron, from one-half of which the 3-inch-wide bar was made.'

Acid Bessemer Steelmaking in South Yorkshire

T. J. LODGE

INTRODUCTION

'For as certain as the age of iron superseded that of bronze, so will the age of steel succeed that of iron.'

Thus spoke Henry Bessemer in 1861 to the Sheffield Meeting of the Institute of Mechanical Engineers when describing his new process for making mild steel in bulk.

His words describing the potential of his new technique received a lukewarm reception from Sheffield's conservative crucible steelmakers. To them, the process appeared to be a non-starter, since early trials at Dowlais and elsewhere had shown Bessemer steel to have severe short-comings.

Despite the initial hiccups, however, Bessemer persevered in the face of mounting opposition from established iron and steel manufacturers. His persistence eventually paid off, and within his lifetime he would see his 1861 prophecy come to fruition.

THE EARLY YEARS

Our story, one of the most incredible from the Victorian industrial era starts in the mid 1850s when Bessemer, the son of an immigrant French refugee, brought to a successful conclusion a series of experiments at St Pancras in London.

Simple enough in themselves – basically blowing air through molten pig iron – the experiments would go on to have far-reaching world wide repercussions, for they signalled the birth of the bulk steelmaking era.

It is a sad fact that for generations the drive to produce better ordnance has led to the development of new metals, and such was the case with Bessemer steel. Bessemer had devised a system of giving cylindrical shells a spin to keep them on course without the need for rifling within the gun barrel, enabling larger shells to be fired with greater accuracy.

The British Government in the form of Woolwich Arsenal pooh-poohed the idea but the French Emperor, Napoleon III, was only too ready to finance Bessemer's work. Bessemer then came up against a physical limitation which prevented existing guns firing his shells; their iron barrels were just not strong enough to resist the discharge. Consequently, his work had to be extended to finding a better metal for ordnance purposes. Huntsman's crucible steel, though widely employed by this time in Sheffield for cutlery and edge tools, was considered too costly for this particular application.

After eighteen months of experimenting to improve the quality of iron, Bessemer hit on the idea of blowing air into molten pig iron in a crucible to oxidise the excess silicon and carbon. The result was a few pounds of soft malleable iron – sufficient to prove the technique, and encourage Bessemer to persevere. He then argued that the use of a refractory lined vessel with holes in the bottom (for blowing in air) would make a greater quantity of iron by the same principle.

When such a vessel was tried the violence of the reaction which took place some minutes after the molten pig was first introduced, literally petrified everyone present. They were equally amazed some minutes later when the violence died down and the material remaining in the vessel was found to be a form of mild steel. Bessemer, describing the incident later, spoke of the reaction as a veritable volcano in a state of active eruption. Bessemer announced his invention to the British Association in Cheltenham in August 1856, his paper being entitled '*The Manufacture of Malleable Iron and Steel without Fuel.*' His work was received with much scepticism – not surprisingly, since to make one ton of crucible steel by then existing techniques consumed some 4 tons of hard coke. Then, to make matters worse, a number of UK ironworks tried the Bessemer technique and reported it to be a failure, producing 'rotten' steel which lacked ductility and broke up during rolling and forging.

It took some little time for the chemists of the day to show that the problem was caused by the unwanted presence of phosphorus and sulphur – two elements found to a greater or lesser extent in the initial pig iron. For his experiments, Bessemer had used pig iron from Blaenavon Ironworks (South Wales), which was relatively low in these impurities, thereby unwittingly guaranteeing success! Very soon the addition of spiegeleisen (ferromanganese) after blowing was found to eliminate hot shortness by rendering the sulphur harmless, and at the same time it reduced porosity in the final product. Thus, provided low phosphorus irons were used the Bessemer process was capable of making good steel.

Because of the bad publicity the Bessemer process received in the mid 1850s there was still considerable prejudice from the 'establishment' within the industry. There was only one option left to Bessemer, and he lacked no convictions when it came to making a decision. With several partners he purchased land on Carlisle Street, Sheffield, erected a works in 1858 and commenced production of Bessemer steel in an area of the country where he was most unwelcome. The business, styled 'Henry Bessemer & Company', was a partnership consisting of Henry Bessemer, his brothers-in-law Robert Longsdon and William Allen, together with William and John Galloway of Manchester. Bessemer's decision to invite the Galloways to join the venture was a wise one, for their old-established business at the Knott Mill Ironworks in Manchester was well versed in engineering matters. Indeed, the firm of Galloway Brothers would soon go on to become a world leader in the supply of Bessemer converters and ancillary equipment – blowing engines, rolling mills and so on.

Bessemer was fully aware of the possible consequences of his actions and later wrote: 'In thus opposing the old-established steel trade in its very midst, we ran the risk of 'rattening', or a bottle of gunpowder in the furnace flues, by which the workmen of Sheffield had earned for themselves an unenviable notoriety'. He was referring of course to the 'Sheffield Outrages', which were the culmination of politically motivated incidents involving restrictive trade practices and associated intimidations.

Well, the local steelmakers might have been a conservative lot, but some of them began to see potential in the process – especially when Bessemer started to undersell them by some £15 per ton and selected users of Bessemer steel reported the product to be good.

With the major problems solved, Bessemer presented a paper to the Institute of Civil Engineers in 1859, emphasising the role his steel could play in the constructional field – building railways, bridges, and so on. This, in fact, was the direction in which his new mild steel industry quickly moved. As a result, the Bessemer process was never a serious threat to the established crucible steel industry in Sheffield; instead the two became somewhat complementary.

THE TECHNOLOGY SPREADS

Even as new outlets for Bessemer steel were opening, Bessemer remained somewhat obsessed with the idea of using it for ordnance. To this end he invited Colonel Eardley Wilmot of the government's Woolwich Arsenal in London to view the process 'in action' at the partners' Sheffield Works. Wilmot was impressed with what he saw, and as he had separate business with John Brown and John Ellis of the neighbouring Atlas Works, he persuaded them to join him in his tour of the Bessemer Works.

Brown and Ellis saw a converter blow and were rapidly 'converted' themselves. This was a crucial moment for Bessemer, for these men were part of the Sheffield 'establishment' who had previously shunned him, and were now publicly acquiescing to the merits of his process. Their Atlas Works (Fig. 1) was one of a new generation of giant iron- and steelworks established on the Don Valley flats east of Sheffield following the opening of the Sheffield and Rotherham Railway in 1838. John Brown & Company was already supplying the burgeoning railways with steel goods – Brown's conical spring buffer, patented in 1848, made him his first fortune – and here was an opportunity to increase business in this lucrative market.

Charles Cammell of the nearby Cyclops Works quickly followed Brown's initiative by also taking out a licence. Cammell's earlier career had been 'on the road', working as a travelling representative for an old established Sheffield engineering tool manufacturer, and he numbered several of the early railway companies, such as the Liverpool and Manchester Railway, amongst his customers. His experiences had left him in no doubt of the business potential that existed for those prepared to supply goods to this 'new-fangled' railway industry.

Brown's Atlas Works was producing Bessemer steel and rolling it to rails by mid-1861; Cyclops was doing likewise before the end of the year.

Acid Bessemer steel was found to be suited to other railway materials – thanks to trials on the London and North Western Railway – and rails, tyres, axles and even boiler plates in Bessemer steel soon became part of the new order.

The next few licences Bessemer granted were to concerns outside Sheffield, and these effectively heralded the spread of steelmaking technology away from the town. The development was more significant than was realised at the time, for it was nothing short of the beginning of the end of Sheffield's centuries-long monopoly in UK steel production.

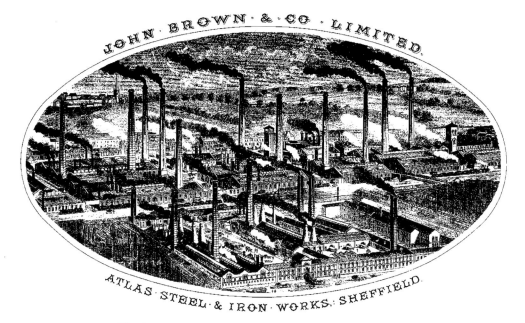

Fig. 1   John Brown's Atlas Steel and Iron Works, Sheffield.

True, two of the new licensees were only just outside Sheffield – Samuel Fox at Stocksbridge and Daniel Adamson at Penistone – but the process accelerated with a vengeance in 1865 when Bessemer converters were installed at the Barrow Iron Works in the Furness district of Lancashire on a site ideally located near an abundance of Cumberland hematite ores. Barrow Works, complete with blast furnaces and Bessemer converters, was also on the coast and thereby unwittingly provided the model for all the world's future 'state-of-the-art' integrated steel plants located alongside deep water ports.

Contemporary references to early Bessemer steelmaking exploits in South Yorkshire are not common. *The Engineer* for 2nd May 1862, however, included an article covering The International Exhibition, in which it was stated that the exhibits provided by Henry Bessemer and Company included 'samples of steel wire, by Fox, of Deepcar'.

In the early days of the mild steel trade, Henry Bessemer had to be very much a salesman to promote his new product. Consequently, he often approached those who were engaged in metal processing, inviting them to try Bessemer metal for themselves so they could see how it compared with their regular feedstock. Having approached Samuel Fox in this fashion, Bessemer got back something of a bonus – mild steel wire which served to illustrate on his exhibition stand yet another use to which his new product could be put.

Processing Bessemer steel to wire must have also made Samuel Fox aware of the potential of this new material, so his decision to acquire a license for Bessemer steelmaking wasn't altogether a surprise.

The installation of Bessemer steelmaking equipment was obviously a large capital under-taking, and not altogether without risk. One historian of Stocksbridge Works, Works Engi-neer Joseph Sheldon, provides us with an interesting insight of the whole business. He claimed that Fox, anxious to get the Bessemer plant erected and working in minimum time, personally handed up bricks to the bricklayers to speed things on! Sheldon also gives us useful information concerning the state-of-the-art technology then employed. The two 5 ton capacity converters and related equipment for Stocksbridge's Bessemer plant were supplied by Galloway Brothers, who were quickly gaining knowledge in the manufacture and operation of Bessemer steelmaking equipment and the necessary ancillaries.

The Bessemer steelworks established at Penistone by a group of partners (Daniel Adamson, George Benson, Robert Braithwaite Benson, Thomas Garnett and William Garnett) is something of an enigma. Daniel Adamson, a Lancashire engineer, was an early advocate of Bessemer steel, which he maintained gave the pressure vessel industry an increased margin of safety over the indifferent wrought iron then widely used for boilers. The foundation stone for the Yorkshire Steel and Iron Works of 'Adamson, Benson, Garnett and Company', as it was known, was laid on Whit Monday 1862 on land acquired by the partners from Thomas Marsden, and the first Bessemer blow made on 1st May 1863. Apart from Bessemer's own works in Sheffield, Penistone proved to be the UK's first true greenfield site development for a Bessemer steelworks.

How much steel plate was produced at Penistone, or was rolled to plate elsewhere from ingots cast there remains a mystery. A crude plan prepared in 1863 shows two buildings – a converter house and an adjacent structure shown as the 'Forge, tyre, plate, rail and bar mill'! The only local publication which we can regard as contemporary – John Wood's *Remarkable Occurrences of Penistone and District* makes no mention of steel plate manufacture, and goes on to state that the plant was sold to Charles Cammell and Company in 1864, 'since which time they have been greatly extended. Bessemer steel, steel rails, tyres, axles, &c. are manufactured.'

The choice of Penistone for the location of the plant may now seem rather strange, but its central position and good rail connections were then important factors. By 1890 the works occupied 25 acres and was turning out 2 000 tons of Bessemer steel products weekly.

As events turned out, the Penistone and Stocksbridge Bessemer plants were very contemporary, and shared a number of similarities. An old diary records that Stocksbridge blew its first charge of Bessemer steel in April 1863 and, using ingots from initial casts, rolled its very first rails on 8th October 1863, to despatch them almost one month later (6th November) to the Great Northern Railway.

THE PROCESS DESCRIBED

The existing steelworks in South Yorkshire which in the 1860's adopted the Bessemer process - Brown's, Cammell's (Cyclops) and Fox's – had all previously relied on traditional crucible melting techniques. Consequently, none operated blast furnaces, which would otherwise have been a valuable source of molten iron feedstock for their newly-installed Bessemer converters. Instead they, and indeed the other Bessemer plants subsequently established in South Yorkshire, were styled along the lines of what later became known as the 'classic' Bessemer layout, based on what Bessemer himself employed at Sheffield. At these plants, bought-in hematite pig iron was melted in coke-fired cupolas and fed alternately to one or other of a pair of converters for blowing down.

A reasonably contemporary description of the actual conversion process, as carried out at Bessemer's own works, appeared in the *Illustrated Guide to Sheffield* published by Pawson and Brailsford in 1879, and is quoted here.

'Bessemer steel is made direct from crude pig iron, the essential part of the process consisting in clearing the iron from its various impurities by subjecting it, while in a liquid state, to the action of small streams of air; an infusion of spiegeleisen being made at the end of the purifying operation in order to give to the steel temper, ductility and certain other necessary properties. The proportion of spiegeleisen used varies according to the temper of steel required, but is much less than the quantity of ordinary pig. The two kinds of pig iron are separately melted in round blast furnaces or cupolas, coke being used for combustion and lime as a flux. A charge varies from four to eight tons, according to the size of the cupola, and is melted in about an hour. In our illustration the cupolas will be seen in the back-ground. In front of them, but on a lower level, is the Bessemer shop, in which is a half-circular pit about four feet deep, with converters at each end (Fig. 2). A converter is a large oval vessel, with a curved nozzle or spout at the top for pouring the metal in and out, and an air chamber at the bottom. It is composed of iron plates, and lined with a fire-resisting plaster made of powdered stone – a hard local stone called 'ganister' being used for the purpose. The air chamber is divided from the caldron of the converter by a thick ganister plate, in which are fire-clay tuyeres full of holes, three-eighths of an inch in diameter, the number of holes in the plate varying from 49 to 185, according to the size of the converter. The air chamber is connected by tubes with powerful blowing engines. The converters rest on strong iron frames, and can be moved at pleasure by hydraulic power.

We enter the Bessemer shop while the pig iron is melting in the cupolas, and note that one of the converters is being heated to a red glow by a fire inside. Suddenly the converter capsizes, the embers falling out, and then rises into a horizontal position, turning its spout to the side of the pit to receive the glowing liquid iron, which streams placidly down an open channel, lined with sand, from the larger cupola. This done, the blast from the blowing engines is turned on, and we are startled with a sudden roar and splutter of flame and sparks, as the blast is forced in strong currents through the apertures of the air chamber into the molten metal. Immediately the converter rises into a vertical position, and we see that the process of converting crude iron into steel of very wonderful qualities has begun in earnest. The blast is turned on before the converter is raised, and is continued without intermission until it falls again, to prevent the metal percolating into the air chamber. Strong pressure is needed to force the air through the liquid metal. Intense combustion immediately follows, the flames from the converter throwing up thousands of explosive sparks, which would form at night a very beautiful pyrotechnic display, if instead of being enclosed by high walls the sparks were thrown into the open air. The blowing goes on until the red glare passes away, and the place is illuminated by a beautiful white light. When the practised eye of the workman in charge sees that the metal is ripe for his purpose, the converter is lowered, and receives the charge from the spiegeleisen cupola. A rapid combination takes place, and the molten iron becomes Bessemer steel. The process of conversion occupies 20 minutes.

Then follows the casting. Between the converters is a hydraulic lift with round stem and oblong top, in one end of which is a vessel called 'the ladle.' The steel is poured from the converter into the ladle, the latter is swung round, and from it the steel descends into a series of moulds in the curve of the pit. To ensure the soundness of the ingots, the moulds are now usually filled from the bottom. The steel solidifies in a few minutes, and the ingots are then removed by means of hydraulic cranes and conveyed, while still red hot, to the rolling mills or hammer shops, where they are made into rails, tyres, axles, piston-rods, or other articles, according to circumstances.

Messrs. Bessemer and Co. have large forges, tyre mills, & c., and manufacture tyres, axles, spindles, piston-rods, and Bessemer steel forgings of all kinds, in which they carry on a large trade.'

Fig. 2   Cammell's Penistone Works Bessemer shop laid out in the classical Bessemer style.

The description dismisses the hot processing which followed casting in a single sentence, but this subject is worthy of a study in its own right. Initially, Bessemer ingots were cogged down using Nasmyth's steam hammer before going for (rail) rolling, but this process was superseded once John Ramsbottom at Crewe introduced a reversing rolling mill in the mid 1860's.

ENTER ALCHEMY

The introduction of Bessemer steelmaking was a real boost for the introduction of steel-works analytical chemistry techniques, and for reasons which were only too apparent.

Classical work by Boyle, Priestley, Lavoisier and Dalton had paved the way for the concept of the Conservation of Matter – a key feature in the development of quantitative chemical analysis. Despite all this, however, the 'new' scientific approach was not held in universal esteem. In particular, Sheffield's crucible steelmakers still preferred their time-tested empirical method for grading steel ingots by estimating 'temper' (actually carbon content) based solely on the crystalline appearance of an ingot's fractured face! They also 'policed' their raw material feedstock by using only high grade Swedish iron - sufficient of a guarantee to ensure low sulphur and phosphorus in the final product.

With the new Bessemer method, something better was needed, since a single blow would produce a cast of several tons in less than half an hour, and accurate and rapid checks on composition became vital. This was equally true of monitoring the quality of the feedstock hematite pig iron, which had to be low in sulphur and phosphorus for a ductile steel product.

Technical matters took a quantum leap forward – both literally and metaphorically – when optical spectroscopy was adopted to study the flames issuing from the Bessemer converter during blowing, and thereby monitor the progressive elimination of silicon, carbon and so on. John Brown & Company pioneered the use of such an instrument, seemingly by the mid 1860s, and it is ironical to think that analytical techniques have come full circle with a vengeance since spectroscopy is now the principal means by which steels and alloys are analysed. Stocksbridge acquired its spectrometer in 1868 for the then princely sum of £5 – roughly the selling price of half a ton (508 kg) of steel rails!

Judging the 'end-point' of a Bessemer blow was a crucial feature of the operation. The blowing period could not be fixed but had to be varied depending on the metalloid (carbon, silicon etc.) content of the hematite iron feedstock. The skilled furnace hand knew when oxidation had reached the desired point by the length, type and colour of the flame issuing from the mouth of the converter. The importance of judging the 'end-point' can be gauged from the fact that in the 1880s (at least) the blower at Stocksbridge was actually paid more than the Bessemer shop foreman, even though technically subordinate to him!

BOOM AND BUST

The success of South Yorkshire's initial Bessemer steel rail producers, coupled with continuing strong prices and high demand, especially from the United States, resulted in the formation in the early 1870s of three further business partnerships in the Sheffield region intent on exploiting the situation. One of these would mark the entry of Rotherham – until this time an 'iron' town – into bulk steel production. The other two were formed with what we can only describe as rather dubious motives.

Sadly, the optimism surrounding these new ventures soon soured, for the boom conditions that spawned them faded away rapidly after American demand fell following the collapse of bankers Jay Cooke in 1873. This set-back was followed by the commissioning of several American Bessemer plants and rail mills, so that even after the financial crisis had been resolved the once lucrative export rail trade to the United States never regained its former status.

In 1871/72 partners Hampton and Radcliffe installed a small Bessemer plant at the Ickles, Rotherham. From an initial modest output of some 300 ingot tons per week, the plant grew at such a pace that by 1879 it was capable of producing 2 750 tons per week, mainly for rail rolling. This achievement is made even more incredible when we realise that the plant was operated by several distinct partnerships during the prevailing unsettled economic climate. In July 1872 the business was floated as the Phoenix Bessemer Steel Co. Ltd. to acquire the adjacent Ickles Rolling Mill but had to be wound up two years later when a bad debt from a failed London iron merchant compounded the company's existing liquidity problem. In November 1875 the receivers sold the plant to Henry Steel, a retired 'Turf Commission

Agent', and a new company – Steel, Tozer & Hampton Ltd. – was registered to carry on the business, with Steel as Managing Director. The name changed in 1883 to the more familiar one of Steel, Peech & Tozer Ltd. following the retirement of director Thomas Hampton. (The remaining directors were Edward Tozer, a well-known Sheffield steelmaker and ex Master Cutler; and Henry Peech, a life-long business associated of Henry Steel). This new company survived the subsequent ups and downs which beset the industry and went on to establish itself as a major player in the railway materials market. Together with associate company Owen & Dyson Ltd. it could proudly and justifiably claim in its adverts to be 'known wherever there are railways.'

The other Bessemer plant established at roughly the same time as that at the Ickles was the works of Brown, Bayley & Dixon at Attercliffe, Sheffield. It was common knowledge in Sheffield that John Brown (by now Sir John) was behind this venture, though it was 'fronted' by his nephew George Brown. Sir John had been 'retired' from the board of John Brown & Co. Ltd. in 1871, and was made to give an undertaking not to set up locally in competition with the Atlas Works. The temptation proved too great, so he adopted the rather transparent solution of involving his nephew to flout this undertaking. The death of George Brown shortly after this, coupled with the drop in trade, put considerable pressure on the company. It survived, however, partly by being one of the first companies to adopt more cost-effective roll cogging following John Ramsbottom's lead at Crewe. (Ironically, John Brown & Co. Ltd. quickly followed this initiative of Brown, Bayley & Dixon, once it was demonstrated that roll cogging dispensed with a large part of the labour force).

The Dronfield works of the Wilson and Cammell Patent Wheel Co. Ltd. is probably the most intriguing of all the Bessemer plants established in the Sheffield region. Registered in March 1872, ostensibly to manufacture railway wheels, the business rapidly became an aggressive participant in the rail trade. The whole saga is dealt with admirably by John Austin and Malcolm Ford in *Steel Town Dronfield* (Scarsdale Publications, 1983). Once the pretext of wheel manufacture was exposed it was apparent that the company posed a major commercial threat to the region's existing rail mills. That was not all. The Dronfield venture was a private partnership set up by members of the Wilson and Cammell families, prominent amongst these being George Wilson and Bernard Cammell (son of Charles Cammell). It was more than sheer coincidence that Charles Cammell was chairman of Charles Cammell & Co. Ltd. – by now a joint stock company – and George Wilson its managing director. Before long shareholders were asking questions about the divided loyalties of Wilson, who was also chairman of the Dronfield firm.

Whatever the truth of these suspicions, the Dronfield complex was a modern, efficient steel plant and rail mill which was able to undercut all its local competitors. It soon established record output levels, both in steelmaking and rolling, and became a recognised world leader in Bessemer steel rail production. In December 1873, *Iron* reported an output of 48 blows in 24 hours, producing 270 tons – 'One of the greatest feats in the history of production of Bessemer steel.'

Ultimately, in 1882, the partners in Wilson, Cammell & Co. Ltd. (the name of the Dronfield firm from 1875) decided to sell the business to Charles Cammell & Co. Ltd. Their reasons were probably threefold – the removal of any further suspicion of 'double-dealing'; the fact that Dronfield could not compete effectively for export orders with more efficient coastal plants which used hot metal feedstock; and the relentless increase in railway carriage rates for raw materials and finished goods, which was significantly eroding profit margins. One year later Charles Cammell & Company adopted a novel approach which solved the last two problems by relocating the Dronfield plant in its entirety to Cumberland alongside a hot metal source. Tragically, the move turned Dronfield almost overnight into a ghost town.

AN ERA ENDS

By the end of the 19th Century Bessemer steelmaking and rail rolling had been relegated to secondary status within the South Yorkshire metallurgical scene. Quite simply, Bessemer steelmaking could be carried out far more economically at coastal sites near hematite ore fields. Many of the former Bessemer converter operators survived by introducing more flexible, cost-effective Siemens open hearth furnaces and diversifying into new products.

Bessemers first licensee, now trading as John Brown & Co. Ltd., was still advertising Bessemer steel in its range of products in 1893 and another, Brown Bayley Steel Works Ltd., continued to do so until some time between 1909 and 1919.

The firm of Henry Bessemer & Co. Ltd. remained very much a 'one-off' of the South Yorkshire Bessemer plants. It never operated a rail rolling mill, concentrating its output on medium and heavy forgings – shafts, axles, crank axles, tyres and other related railway materials. Sometime in the 1890s it gave up its very 'raison d'être' – making steel by the Bessemer process – and installed acid open hearth furnaces. It survived until the mid 1920s Depression, and after closure the goodwill of the business was acquired by John Baker & Co. (Rotherham) 1920 Ltd., a firm also specialising in the manufacture of railway materials. The Baker family members on the board of directors were delighted to gain control of such a prestigious name in steel, and changed the name of the business to John Baker & Bessemer Ltd. to reflect its new status. This successor company remained an important supplier in the UK's railway materials sector until it ceased trading in the mid 1960s, when it in turn was bought out for its order book by its two large rivals, the English Steel Corporation Ltd. and the United Steel Companies Ltd.

There is a particular poignant aside to Bessemer's Sheffield business. The firm was initially managed by Bessemer's brother-in-law William D. Allen, and he was succeeded by his son Harry Allen when the original partners retired and the business was floated on the Stock Exchange. Harry became an active member of Sheffield's steel manufacturing 'establishment' – the very clan which had once so soundly re-buffed Henry Bessemer – and by 1899 he was serving as Senior Warden in the Cutlers' Company of Hallamshire, a Livery Guild organisation which had safeguarded the status of the region's metal manufacturers since the 17th Century. Literally only months before he was due to be elected Master Cutler – the figurehead of all Sheffield's steel industries – Harry succumbed to influenza which turned to pneumonia and he died on 23rd February 1899 aged 45. Metallurgist and industrialist Robert Hadfield, as next in line, had the sad duty of taking up office as Master Cutler somewhat earlier than he had anticipated. One cannot help but think that had Harry Allen become Master Cutler, and had Henry Bessemer lived a little longer to see it, the latter might well have regarded the occasion as the ultimate recognition of his process and his faith in it.

I have been unable to ascertain when the Bessemers at Cammell's Cyclops works were closed down. Cammell's progressively concentrated its South Yorkshire acid Bessemer operations at its Penistone Works, and it is likely that this rationalisation was completed by about 1900.

At the turn of the century the only acid Bessemer plants in South Yorkshire with a reasonably assured future were those at Penistone, Rotherham and Stocksbridge. By this time all three were members of an Inland Rail Makers' Association, which ensured their mills continued to receive rail orders from British railway companies on an unofficial rota basis. (The following chapter deals with the Association's activities in more detail).

The Great War put the steel industry under tremendous pressure. To increase output of steel billets (for shell manufacture) Stocksbridge closed its Bessemer shop so that the adjacent rolling mill could be enlarged. The works continued to roll acid Bessemer rails, however, using ingots supplied by Steel, Peech & Tozer Ltd. under a special agreement. This co-operation on ingot supply proved to be the first tangible move which ultimately brought the two firms together to form the nucleus of the United Steel Companies.

The Bessemer plants at Penistone and Rotherham survived as working units into the 1920s, though each by then was complemented with open hearth melting units. It was inevitable, once the post war Depression began to bite, that the days of Bessemer steelmaking in South Yorkshire were numbered. An era ended when the region's last converters were finally silenced in the late 1920s.

THE LAST LINK

The last acid converters in use in the UK – the two 30 ton capacity vessels at British Steel's Workington Works – were effectively direct descendants of those in the Sheffield plants of the 19th Century. It was indeed fitting that one of the Workington converters was saved when its working life was over, and now graces in a most impressive manner the entrance to Sheffield's Kelham Island Industrial Museum complex (Fig. 3), less than a mile distant from the site of the UK's first commercial Bessemer steelplant.

BIBLIOGRAPHY AND ACKNOWLEDGEMENTS

The two most comprehensive works dealing with the acid Bessemer process in South Yorkshire are 'The Sheffield Rail Trade 1861–1930' by K. Warren *(Trans. Institute of British Geographers,* June 1964) and *Steelmaking: 1850–1900* by K. C. Barraclough (The Institute of Metals, 1990). The former is a concise economic history, while the latter deals at length with the industry from a technologist's point of view.

In the light of this, and the fact that Dr Barraclough's work is readily available, I have slanted the present offering to include relevant information which neither of these works address adequately. A certain amount of repetition was, however, necessary in order to reiterate the broader chronology of the process. Readers will gain a greater appreciation of the

Fig. 3   Acid Bessemer converter ex Workington Works, shortly after its installation at the entrance to Sheffield's Kelham Island Museum.

full picture by reading this section in conjunction with that on South Yorkshire's Steel Rail Trade, bearing in mind that rails were the 'force majeure' in the region's Bessemer trades.

I am grateful to three colleagues of long standing – Ron Dyal, Gordon Green and Ken Plant – for their generous assistance with the provision of historical information.

Appendix

Acid Bessemer Plants in South Yorkshire
Number/Size of converters

	1867[1]	1871[2]	1878[3]	1881[4]	1924[5]
Bessemer & Co.	1 X 1/2t 2 X 4t	1 X 1t 2 X 3 1/2t 2 X 5t	2 X 3t 2 X 5t	2 X 3t 2 X 5t	–
Brown (Atlas)	4 X 5t 2 X 11t	2 X 7 1/2t 2 X 10t 2 X 15t	4 X 7t 2 X 10t	4 X 7 1/2t 2 X 10t	–
Cammell (Cyclops)	2 X 4t	4 X 5t	4 X 4t	4 X 4t	–
Penistone	4 X 7t	4 X 5t	2 X 5t 2 X 7t	2 X 5t 2 X 7t	2 X 13t
Fox	2 X 4t	2 X 5t	2 X 5t	2 X 5t	–
Ickles	Not built	2 X 5t	2 X 2 1/2t 2 X 6t 2 X 7t	2 X 2t 2 X 6t 2 X 8t	2 X 15t
Brown, Bayley	Not built	2 X 5t	4 X 8t	2 X 4t 4 X 8t	–
Dronfield	Not built	Not built	4 X 6t	–	–

1. *Revue de l'Industrie du Fer en 1867* by S. Jordan, vol. 4. Paris, 1871.
2. Collected Statistics, *Journal of the Iron and Steel Institute*, 1871.
3. *Steel, Its History, Manufacture, Properties and Uses*, by J.S. Jeans, Published by Spon & Co., 1880.
4. *The Coal and Iron Industries of the United Kingdom*, by Richard Meade, Published by Crosby, Lockwood & Son, 1882.
5. *Ryland's Directory of the Coal, Iron, Steel, Tinplate (etc) Trades*, 1924 Edition.

The capacities quoted need to be interpreted with a little care. The figures for John Brown & Co. (Atlas) in 1867, for example, were actually quoted as four converters at 4 ton–6 ton capacity, and two converters at 10 ton–12 ton capacity. Most of the tonnages quoted appear to be nominal rather than absolute figures. On this basis the four large converters at Bessemer & Co. are likely to be effectively the same vessels throughout. A similar argument applies with the Ickles converters (apart from the 1924 listing).

South Yorkshire's
Steel Rail Trade

T. J. LODGE

INTRODUCTION

Before Bessemer steel was available, the world's expanding railway networks were quite literally 'iron roads', laid exclusively with wrought iron rails.

The vast bulk of this trade, both national and international, was supplied by a dozen or so works established at the heads of the valleys of industrial South Wales. True, there were isolated iron rail makers elsewhere, and in South Yorkshire the Park Gate Iron Company – a predecessor of the Rotherham complex of British Steel Engineering Steels – was the most important of these. However, Bessemer's converter was destined to change the established rail trade in a most profound manner, with far-reaching implications for both South Wales and South Yorkshire.

Yet Bessemer tended to be rather dismissive about the steel rail – the one product which ultimately gave his process its greatest commercial success. In Chapter 14 of his *Autobiography* he even goes so far as to say that the 'adoption of my process by the ironmaster for making rails went far to discredit it.' His concern was very real, since in view of the controversy surrounding the introduction of his process, he was now most anxious to distance it from anything which might associate it with inferior products.

Rails at that time had a bad press. Abram Hewitt, the influential American ironmaster, voiced a particularly damning opinion in his work on 'The production of iron and steel in its economic and social relations' which appeared in the *Report of the US Commissioner to the Universal Exposition, Paris, 1867*. Abram admitted that 'it was humiliating to find that the vilest trash which could be dignified by the name of iron went universally by the name of the American rail.'

Once Sheffield steelmakers adopted the process, however, a whole new scenario was created. Sheffield had long enjoyed a reputation for quality in the American colonies, and this went a long way to establishing a thriving export trade by dispelling the old fears about wrought iron rails. The commercial steel rail trade which subsequently developed had its roots firmly in the Sheffield region, which makes this chapter particularly relevant to our story.

PROVING THE PRODUCT

At this point it is appropriate to deal with the several claims made by different authorities for steel rail 'firsts'.

Steel rails soon became a success in their own right, so we shouldn't be surprised that the various protagonists each became keen to claim (mostly retrospectively!) a share of the glory attached to 'the first.'

Edward P. Martin of the Dowlais Iron Company, in his Presidential Address to the Iron and Steel Institute in 1897, recounted some of this early history. Dowlais took out an early Bessemer license, and Martin says that 'it was at their works that Bessemer steel was first rolled into rails'. Bessemer's *Autobiography* informs us the rails were rolled on 6th September 1856 from two 10 inch square ingots cast at Bessemer's experimental works at Baxter House in St. Pancras, London. The material had been converted from Blaenavon pig iron without the addition of spiegel or ferro-manganese. Martin later had a piece analysed at 0.08% carbon, 0.16% sulphur and 0.43% phosphorus, with traces of silicon and manganese. In other words, Bessemer had produced a high carbon wrought iron in this instance rather than a true steel. Later attempts to produce good Bessemer steel consistently at Dowlais were not successful, and as a result, commercial rollings of steel rails did not begin here until 1864.

In 1857 Robert Mushet succeeded in crucible melting part of an ingot made from an abortive Bessemer blow at Ebbw Vale in which the 'steel' had been 'burnt,' or over-oxidised. He then deoxidised this re-melt with ferro-manganese, had a new ingot cast and rolled to a rail which was laid at Derby station for trials. This hardly counts as a true 'Bessemer' steel rail and the unorthodox production route can in no way be regarded as commercially viable. Even so, the rail – technically the first steel rail to see extended use – is said to have lasted in service until 1873.

The next to stake a claim was J. D. Ellis, the partner of John Brown at the Atlas Works in Sheffield. Sir Allan Grant, author of *Steel and Ships* (a history of John Brown & Co. Ltd.) claimed there was a legend to the effect that Ellis, 'one dark and stormy night', took a gang of his own platelayers to surreptitiously remove one of the iron rails of the Midland Railway line which was adjacent the Atlas Works, and replace it with a Bessemer steel rail. Only months later, when the railway company's engineer noticed its superior wearing qualities, did Ellis admit the secret. We are not given a date, but the episode must be about 1861, the year Brown's began manufacturing steel rails. I have been unable to substantiate the claim with contemporary material, but irrespective of this tale it can be said that Brown's were the first commercial producers of steel rails.

Bessemer himself next enters the 'competition', and tells us in his *Autobiography* how he supplied steel blooms to the London and North Western Railway (Britain's largest) for rolling through the iron rail mill at its Crewe Workshops complex. It was a particular triumph for Bessemer, for not much earlier than this he had been soundly rebuffed by the LNWR's Chief Mechanical Engineer, John Ramsbottom, when he suggested that Bessemer steel rails should be tried. Ramsbottom had countered 'Mr. Bessemer, would you have me tried for manslaughter?' showing just how strong was the suspicion regarding Bessemer steel integrity following the early failures at Dowlais and elsewhere.

Bessemer steel rails were laid in November 1861 at Crewe Station and in 1862 at Chalk Farm on the main line out of London (Euston). The latter must count as the first authenticated occasion when a proper trial on Bessemer steel rails was conducted, for those laid at Crewe were soon lifted and put on display at the 1862 London International Exhibition. The vastly superior wearing qualities of the trial rails at Chalk Farm guaranteed the future of the steel rail trade.

John Brown, whose Atlas Works was on Carlisle Street in Sheffield, was Bessemer's first licensee for the process in Sheffield, and with the success of steel rails assured, it wasn't long before others in the area – Charles Cammell & Company at the nearby Cyclops Works, and Samuel Fox at Stocksbridge – licensed the Bessemer process for rail production.

At this point the whole business mushroomed. By the late 1860s South Yorkshire's five Bessemer plants were producing steel and rolling it to rails at such a rate that they were responsible for half the UK output of steel rails. Market penetration, particularly in America, was soon achieved, helped largely by Sheffield's reputation for quality and also the rather indifferent quality of the wrought iron rails supplied by South Wales works. This buoyant overseas demand brought about further expansion of the trade, and by 1875 the number of local producers had grown to seven, with an estimated total annual output of some 250 000 tons of rails. This was about one-third of the national production, the balance being made up by Bessemer works established elsewhere in the meantime, notably in South Wales and Lancashire.

ASPECTS OF COMMERCIAL PRODUCTION

The surviving letter copy books of Stocksbridge steelmaker Samuel Fox for the 1860s and 1870's provide a valuable insight into the formative years of the steel rail trade – its magnitude, customers, economics; even its technical problems. Though some of the problems highlighted were unique to Stocksbridge, most were shared by all South Yorkshire's producers of Bessemer steel rails, which makes the information of special interest.

Fox established his business at Stocksbridge on the Little Don River nine miles northwest of Sheffield in 1841/42, drawing steel wire and fabricating related products in a water powered mill originally erected in 1794 for spinning cotton. Much of the initial output went into the making of hackle and gill pins for the West Yorkshire textile industries. Diversification into steel umbrella frame and crinoline skirt wire manufacture followed, and it is reputed that Fox's 'Paragon' umbrella frame patent of 1852 ultimately earned him £1/2 million.

Capital of such magnitude enabled Fox to expand into bulk steel production. In 1862 he took out one of the first licences for Bessemer's process and erected a steelplant housing two 5 ton capacity converters immediately upstream of his wire manufactory. (The raw material stocking area for the plant, incidentally, is still referred to as 'Bessemer Top', though the last converters here were silenced eighty years ago.)

Rail manufacture at Stocksbridge began in 1863 using a two stage forming process – cast steel ingots from the Bessemer plant were first cogged into blooms under the steam hammer then rolled to the required section in a steam driven finishing mill. Rails were typically produced in 15ft–25ft lengths up to 80lb/yard section.

Rails were first recorded as a product at Stocksbridge in the Sales Ledger for December 1863. In that month an entry appears for £1369–14s–10d (£1369·62) for rails supplied to an unspecified home customer. The customer was the Great Northern Railway (GNR); and the quantity almost certainly 150 tons. Stocksbridge was now well and truly in the tonnage league.

Later, Fox would claim to have supplied several thousand tons of steel rails to the GNR between 1863 and 1869. He often quoted the fact as a 'testimonial' to new customers who were seeking assurance on the quality of his product. Rails were also supplied in this early period to the Edinburgh and Glasgow Railway (later the North British Railway); the North London Railway; Midland Railway; North Eastern Railway (NER); and Manchester, Sheffield and Lincolnshire Railway (MSLR).

By the late 1860s the GNR was still Fox's most important rail customer. During 1868, for example, 1200 tons were supplied to the GNR, with a further 400 tons to the NER, at prices between £11 and £12 per ton. Other customers now included the Cambrian Railways, the Cheshire Lines Committee (CLC) and the London, Brighton & South Coast Railway. India and Australia were favourite overseas destinations, and rail enquiries were received (via agents) from Russia and America but Fox complained that because of very competitive prices he could not take on this latter business with the usual agent's commission.

Later, difficulties with access to suitable raw materials and increased carriage rates for overseas orders (once the home market became saturated) put the Sheffield rail manufacturers at a real disadvantage to those located on the coast. This difference was further heightened by the fact that the coastal plants in question, following pioneering work in Cumberland, fed their Bessemer converters with more cost-effective liquid iron from blast furnaces; so-called 'hot metal'. By contrast, Sheffield's Bessemer plants were all 'cold metal' charged, having to employ expensive coke to remelt the pig iron in cupolas before it could be fed to the converters.

Thus even as new local Bessemer works – notably the Ickles Bessemer Works (Rotherham) and Brown, Bayley & Dixon – entered the rail business in the early 1870s the trade was nearing its zenith for South Yorkshire (Fig. 1). This was even more true for the wrought iron rail trade, though it was something of a coincidence that the big local producer – the Park Gate Iron Company Ltd. – ceased rolling iron rails in 1876.

The interest of steelmakers such as Fox in the Rail Trade was only to be expected. In many ways the infant Bessemer industry opened up a market for steel rails the potential of which can be likened to the boom situation which came a century later with the discovery of North Sea gas and oil.

Fox went into the ultra novel business of Bessemer steelmaking and rail rolling with typical thoroughness. His choice of Richard Price-Williams as first Manager of the Stocksbridge Bessemer plant and Rail Mill would have been difficult to better. Price-Williams had been appointed Manager of Bessemer's trial steelworks at Greenwich, London, and is said to have been involved in perfecting the technique somewhat earlier. Furthermore, he had later trained as a civil engineer on the London & North Western Railway, where he had been encouraged to investigate the use of steel rails.

Recalling the period later in life, Price-Williams claimed that the performance of steel

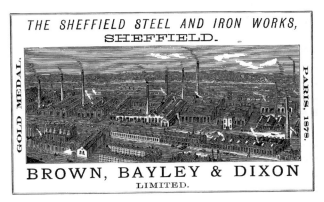

Fig. 1 Two page advert, which appeared in the 1879 edition of Pawson & Brailsford's *Illustrated Guide to Sheffield*, was one of very few run by Sheffield's Bessemer steelmakers specifically mention Bessemer steel rails as a product.

rails so impressed him that he gave up his position on the railway to join Samuel Fox 'in erecting the Bessemer plant and rolling-mills at Stocksbridge'. He was given a surprisingly free hand by Fox, whose autocratic nature on this occasion took second place to his business sense. Under the circumstances, one is inclined to suspect that Price-Williams came to Stocksbridge on Bessemer's personal recommendation to Fox. Such a move would have benefited both Fox and Bessemer. It would have materially assisted Stocksbridge to come to grips with the new technology and at one and the same time minimised the chance of the venture failing, something which would have reflected badly on the newly-established credibility of Bessemer's process, to the detriment of further licensing agreements.

One of Price-William's first decisions in his new capacity took him to the Dowlais Iron Company's rail mill in South Wales in the company of Fox's Wire Mill Manager. The General Manager at Dowlais, William Menelaus, afforded the pair every opportunity to witness the rolling of iron rails at first hand. As a result, said Price-Williams, *'the mill we constructed at Stocksbridge was on the lines of Dowlais but, of course, very much increased in strength to deal with the much tougher material'.*

'Success breeds success', so the saying goes, and Price-Williams played no small role in guaranteeing Stocksbridge's success in the Rail Trade. His paper *'On the Maintenance and Renewal of Permanent Way'*, which drew heavily on the performance of Stocksbridge rails on the Great Northern Railway, earned him the Institute of Civil Engineer's Telford Gold Medal in 1866. A Watt Gold Medal from the Institute of Mechanical Engineers followed four years later for a related paper dealing with the use of steel in railway rolling stock. Perhaps the pinnacle of his career came in 1898 when he was awarded, somewhat retrospectively, the Iron and Steel Institute's Bessemer Gold Medal for *'the active part he took in the early days of the use of steel on the railways when others had not the courage to use it'*. The words were Bessemer's own, and the occasion must have been a particularly moving one for Price-Williams, for they were penned by the great man from his sick bed a mere three weeks before his death.

For Fox, inadequate transport for raw materials and finished product was a real dilemma. His eventual solution – promoting and building the Stocksbridge Railway – is worthy of a study in its own right. Suffice to say here that prior to the line opening in 1878 all Fox's incoming pig iron feedstock and outgoing rail products had to be moved by horse and cart the 1 1/2 miles between Stocksbridge and the MSLR railhead at Deepcar. Fox suffered this situation for fifteen long years at the height of the rail boom when his South Yorkshire competitors – without exception – were rail served from adjacent main lines. Small wonder that in 1869, when he wrote to tell the CLC that he was not able to supply more than 200 tons of rails per month, he bitterly remarked that he 'would be doing thousands of tons, rather than hundreds' were his works rail linked. Significantly, one of the railway companies which formed the CLC – the MSLR – had actively blocked Fox's efforts to get his branch railway from Deepcar to Stocksbridge.

Fox's total output at this time can be gauged from his consumption of pig iron – some 300 tons per week in 1873. Allowing for conversion losses during processing, this equates to something like 13 000 tons per annum of steel rails.

Despite precautions taken with the choice of pig iron feedstock, a number of rail breakages reported by the GNR to Fox in 1867 gave some cause for concern. Casts with unusually high phosphorus are almost certain to have been involved, and it may be significant that from about this time Fox began to employ the services of William Baker, a Sheffield-based analytical chemist.

On the marketing front, occasionally rather blatant attempts to 'woo' customers were made, with a view to stabilising the order book. In March 1872, for example, Fox wrote to Francis W Webb, successor to John Ramsbottom as Chief Mechanical Engineer of the London and North Western Railway, offering him shares in the newly floated public company of Samuel Fox & Co. Ltd. as he wished Webb 'to take an interest in the business'! Generally, however, Fox was more open in his business dealings, and it has to be remembered that such actions were not considered inappropriate at this time anyway. Webb, incidentally, had been a pupil at Crewe, assisting Ramsbottom with the layout of the Bessemer plant there, before breaking his railway service to gain practical experience in Bessemer steelmaking at the Bolton Iron & Steel Co. Ltd., a company set up by the Hick and Hargreave families to produce Bessemer steel products for the railway companies and others.

To remain in business supplying the very competitive rail market Fox appreciated he must reduce his manufacturing costs. He therefore considered two schemes to create an integrated works where molten iron straight from the blast furnace was taken to an adjacent Bessemer plant, but was unable to implement either.

Writing to his London agent in October 1876, Fox voiced the fears of all those connected with the local rail trade who realised that its days were numbered. 'It becomes more manifest every day that unless the Railway Companies interested in this district will come to the help of the Rail Manufacturers the trade must go elsewhere and these Railway Companies will suffer along with the manufacturers. If we are to compete with other districts we must have reduced carriage rates for pig metal from Cumberland and for rails to Liverpool, London, Manchester and elsewhere'. He added, somewhat prophetically, 'if the trade once leaves it will be impossible to bring it back'. Shortly after, he learned that Wilson Cammell's Dronfield plant had enjoyed reduced carriage rates to Ayrshire from the Midland Railway – 10/- (50p) per ton against 20/- (£1.00) quoted to Fox for the same destination. His anger is best left imagined.

Retaining a share in the available rail business became an even bigger problem for South Yorkshire's rail makers when export orders assumed prominence. Fox, for example, was telling his London agent in 1877 that 'English orders yield no profit' yet 'There seems to be no prospect of orders from either Russia or America for a long time to come.' Hardly surprising: at the same time Fox informed a fellow director that 'We have just lost two contracts, the orders having been taken at £6 by Wilson & Cammell'. (The price was per ton, exclusive of delivery, and virtually half what Fox's charged.)

THE RAIL MAKERS' ASSOCIATION

As competition from the coastal Bessemer plants at Barrow, Workington and Middlesbrough intensified, particularly on export orders, the Sheffield railmakers reacted by forming a cartel to stabilise the market by fixing prices and allocating orders.

Price fixing was nothing new to the Sheffield steel trades. By the late 18th century several edge and hand tool makers were already working to standard price lists, following agreements for fixed earnings with the various trade unions involved in the actual manufacture. And much earlier than this – in the 1660s to be precise – attempts had been made by local ironmasters to fix prices.

The first documentary evidence I have found of a local rail cartel dates back to the early 1880s, when Samuel Fox wrote to D L Schönberg, his London agent, outlining his thoughts on how such a cartel could operate. Fox proposed that each member entered into a bond of £10 000 not to undersell each other. Any member breaking the pledge would have his bond confiscated as a fine and divided amongst the other manufacturers equally.

His proposals seemed to find favour with others in the trade, and a 'Local Railmakers' Syndicate' was formed between Fox's, Cammell's and Steel, Peech & Tozer. Orders were shared out on the basis of plant capacity, giving Fox's a 1/5th share and the other two partners 2/5ths each. The syndicate was based on the classical German Kartel system. Each member lodged a bond of ten promissory notes of £1000 each with a neutral body (a solicitor acting as secretary for the syndicate) and agreed not to undercut the mutually agreed prices when tendering for contracts. Any breach of the agreement resulted in the offending member forfeiting all or part of the bond, the money being shared between the syndicate's other members.

Complaints on market share were the biggest threat to the smooth running of the cartel. When, for example, Fox's installed a new rail mill in 1891 the company claimed it was entitled to a greater share of the available orders received by the Inland (ie Sheffield area) Steel Rail Makers' Association, on the basis of the mill's higher capacity. Fox's request was refuted most strongly by Cammell's of Sheffield and Penistone, who had claim to 40% of the Inland Association's allocated tonnage. Ironically, it later transpired that for all its indignation over the matter, Cammell's was actually flouting the association's rules by accepting orders 'on the side'. Not declaring this business to the association allowed Cammell's to exceed its quota without forfeiting for the excess tonnage. A case of 'do as I say', and not 'do as I do'.

The tenuous nature of cartels meant that each was only able to exert a short-lived effect on the market, and this was certainly the case with the Inland Steel Rail Makers' Association. Even the adoption of international iron and steel cartels in the 1920s, for example, was eventually found insufficient to deal with the world's deep rooted economic crisis (Fig. 2).

RAIL SPECIFICATIONS

Prior to the introduction of the steel rail, the only procurement specifications rigorously exercised in the iron and steel trades were those for ordnance and armour by the Admiralty Board and the Army. These were invariably on the basis of 'fitness for purpose', with proving trials to verify this.

Railway companies were the first really large companies to evolve in the Victorian capitalist system, and had quickly to develop management systems for efficient day to day operations, forward planning and so on. Supplies departments were established, and before long procurement specialists began to define requirements more clearly than the vague 'fitness for purpose' clause. Rails were some of the first steel products for which more precise specifications were evolved, i.e. the concept of material within a range of chemical compositions and capable of meeting specified minima with regard to physical properties (strength, ductility etc.).

The railway companies efforts to achieve this desirable state of affairs was not always appreciated by their suppliers. There is a delightful account of an Inland Railmakers' Association Meeting for 27th February 1895 which discussed fixing a price for a Great Northern Railway enquiry. It was revealed that an analytical specification had been introduced by the railway company based on composition checks made on a 20 year old Bolckow rail which had displayed exceptional wear-resisting characteristics. The RMA's members were most indignant at this development, and 'unanimously agreed that if railway companies were going to begin this sort of thing, that they would have to pay for it.' The customer/supplier relationship was obviously not at the sort of level that we strive to achieve nowadays.

Matters, however, came to a head when an express passenger train was derailed on a broken rail at St. Neots on the Great Northern Railway on 10th November 1895. At the

Fig. 2   View of the finishing stands of Steel, Peech & Tozer's Ickles rail mill in the 1920s.

instigation of the Board of Trade, Thomas Andrews of Wortley Iron Works, near Sheffield, carried out a wide ranging examination of the broken rail, along with comparison pieces taken from serviceable rails. Andrews, a self-styled 'Consulting Metallurgical Engineer and Chemist,' reported on chemical composition, mechanical properties and metallurgical structures. His results, which particularly highlighted deterioration by fatigue, were published *in extenso* in *Engineering* through 1897 and 1898. From now on no one would argue against the desirability of such tests.

FINALE

The South Yorkshire railmakers last major home order for rails was a massive 35 000 tons supplied for the building of the last main line into London – the MSLR's extension from Nottingham, begun in 1894. Let at £4/7/6d (£4–37 1/2p) per ton, the order was shared between Cammell's Penistone Works, Samuel Fox's and Steel, Peech and Tozer.

By the turn of the present century the rail trade in South Yorkshire was very much secondary to other steel products, though rail rolling was to survive until the late 1920s at

Stocksbridge, the Ickles (Rotherham) and Penistone. At Fox's, on a relatively new mill dating from 1891, conventional rail sections were rolled to the end, but Steel, Peech & Tozer Ltd. used the Ickles rail mill latterly to produce rather limited tonnages of special rail sections for street tramways. The formation of the United Steel Companies after the 1914–18 War – initially by a merger of Steel, Peech & Tozer with Fox's – was the determining factor which cut short the lingering death of the local rail trade, for cold economics dictated that rail rolling be carried out at only one USC site, and Workington Works was chosen as the most suitable. The choice was a wise one – today British Steel's Workington Works is the sole producer of rail sections in the UK, and exports still count for a large proportion of the business.

Coincidentally, the Workington Works too has some of its roots in the nineteenth century Sheffield district rail trade, being formed in 1909 by the merging of two adjacent plants – the works of the Moss Bay Hematite Iron & Steel Company and the Derwent Works of Cammell, Laird & Co. Ltd. One of these, the Derwent Works, entered the rail market with a second hand Bessemer steel plant and rail mill complex, both of which came from Dronfield, near Sheffield. The transfer of the Dronfield Bessemer shop and rail mill complex to Workington in 1883 was a prophetic one, for it signalled the beginning of the end for the Sheffield rail trade – an end which closed a major chapter in Bessemer steelmaking.

The demise of the rail mills at Stocksbridge, Penistone and Rotherham in the Depression years following the Great War brought to an end an era: a period spanning six decades during which time Sheffield rose from a cutlery and edge tool producer through the nation's largest roller of steel railway rails to become one of the world's largest centres of armament and ordnance manufacture. The Bessemer process played the significant 'mid-field' role in this important transition from light to heavy trades, and in consequence Sheffield owes it a great deal.

EPILOGUE

Bessemer's initial fears about steel rails would prove groundless, and in his lifetime the steel rail would become the Bessemer product which arguably had the greatest effect on the civilised world. Bessemer eventually acknowledged this when he prepared a synopsis of the Bessemer steel industry which appeared in *Engineering Review* for 20th July 1894. 'This new material,' he wrote, 'has covered with a network of steel rails the surface of every country in Europe, and in America alone there are no less than 175 000 miles of Bessemer steel rails binding together its widely-scattered cities.'

Charles Mackay gave a more romantic description of the profound social effect of railways in a poem published in the *Railway Examiner* for 25th November 1845.

> *'Lay down your rails ye nations near and far,*
> *Yoke your full trains to steam's triumphant car,*
> *Link town to town and in these iron bands,*
> *Unite the estranged and oft embattled lands.'*

Mackay wrote of an earlier Railway Age, but his sentiments were if anything even more applicable once Bessemer's steel bands replaced those of iron.

ACKNOWLEDGEMENTS

The majority of this chapter is based on information in the letter copy books of Samuel Fox & Co. Ltd. The other information relating to Samuel Fox & Co. Ltd.; Steel, Peech and Tozer Ltd.; and the United Steel Companies Ltd. is from the minute books and other official records of these companies. All these documents are part of the archives of British Steel plc, and I am grateful to the company for access to this material.

Information concerning Price-Williams is a summary of that which appeared in the *Journal of the Iron and Steel Institute,* notably on his award of the Bessemer Gold Medal in 1898 and his Obituary Notice (*JISI 1916*).

The Bessemer Process in the North West of England

C. BODSWORTH

There are few visible signs remaining to indicate that this part of the country was once the home of a thriving iron and steel industry. Yet in 1878 more than 40% of the 112 converters installed in Britain were located in the North West (see Fig. 1) and in 1889, when Bessemer steel production reached a peak of 2 175 000 ingot tonnes, approximately one sixth of the total was produced in the North West of England. Furthermore, the acid converter process was operated continuously here for over 100 years, until the last heats were blown in 1974.

LANCASHIRE SITES

Henry Bessemer at various times contracted the firm of Galloway Brothers, engineers, at Knott Mill Iron Works in Manchester to manufacture and supply equipment which he had designed for the exploitation of his numerous inventions. It is evident that he was in contact with the company when he started his steel making experiments, for Galloways obtained the first licence to manufacture malleable iron by the converter process, The licence, which gave sole rights to use the process in Manchester and ten miles around, was granted before Bessemer made the first public disclosure of his work in his address to the Cheltenham meeting of the British Association.[1] A small fixed converter was erected at the plant but, in common with the experience at the Dowlais and Govan works, the trials were disastrous. When Bessemer realised that this was caused by the inability of the acid process to remove the high concentrations of sulphur and phosphorus from British pig iron, he produced ingots of carbon steel from purer Swedish pig iron in his converter at his St. Pancras premises in London. The ingots were forged into bars which were supplied to the craftsmen at the Galloway works. In use, they were unable to distinguish the Bessemer bars from the normal (crucible) steel supplies.[2]

Bessemer now sought a suitable supply of British pig iron which would be less expensive and more readily available than the Swedish charcoal pig iron. The most abundant supplies of hematite iron ore in Britain were located in Cumberland, but the pig iron which was produced locally from the ores contained too much phosphorus for refining in the acid converter process. During a visit to the Workington Iron Company in West Cumberland Bessemer noticed that fine hematite ore, which had been used for fettling the puddling furnaces in Staffordshire, was being returned to the iron works in the form of tap cinder slag which was then added as a flux to the blast furnace charges. The tap cinder had absorbed phosphorus

Name and Situation of Works.	No. of Converters	Capacity of Converters.
		tons cwt.
Henry Bessemer and Co., Limited, Sheffield	2	3 0
	2	5 0
Bolckow, Vaughan, and Co., Limited	4	8 0
John Brown and Co., Limited, Sheffield	4	7 0
	2	10 0
Sheffield Steel and Iron Works, Attercliffe, near Sheffield. Brown, Bayley, and Dixon, Limited	4	8 0
Charles Cammell and Co., Limited, Sheffield, Cyclops	4	4 0
,, ,, Yorkshire	2	5 0
	2	7 0
Wilson, Cammell, and Co., Dronfield	4	6 0
Weardale Iron Company, Tudhoe, Ferryhill	4	2 10
The Glasgow Bessemer Steel Company, Limited, Atlas Works, Glasgow	2	3 0
Samuel Fox and Co., Stockbridge Works, Sheffield	2	5 0
Bolton Iron and Steel Works, Bolton	4	6 0
London and North Western Railway, Crewe	2	3 0
Manchester, Sheffield, and Lincolnshire Railway Company, Gorton Works, near Manchester	2	3 10
Mersey Steel and Iron Works, Liverpool	10	5 0
Manchester Steel and Railway Plant Company, Gibraltar Works, Newton Heath, Manchester	4	3 0
Barrow Hæmatite Steel Company, Barrow	16	6 0
The Dowlais Iron Company, Dowlais	2	7 10
	2	6 0
	2	5 0
Ebbw Vale Company, Ebbw Vale	4	6 0
	2	8 0
West Cumberland Iron and Steel Company, Workington	2	5 0
	2	8 0
Phœnix Bessemer Steel Works, Rotherham. Steel, Tozer, and Hampton, Limited	2	7 0
	2	6 0
	2	2 10
Carnforth Hæmatite Iron Company, Limited	2	6 0
Patent Shaft and Axletree Company, Wednesbury	4	3 0
The Moss Bay Hæmatite Iron and Steel Company, Workington	2	7 0
Rhymney Iron Company, Limited. Rhymney	3	8 0
Blaenavon Iron Company, South Wales	2	3 0

Fig. 1   A complete list of works engaged in the manufacture of Bessemer steel in the UK in 1878.[5]

from the iron in the puddling furnace and this was transferred to the pig iron during the smelting operation. When the cinder was deleted from the blast furnace charge the Cumberland pig iron proved to be eminently suitable for the acid Bessemer process. This became the primary source of converter melting stock until towards the end of the 19th century when Spanish hematite ores were imported in bulk quantities.

The quality of Bessemer steel had been proven by the Galloway trials and a supply of suitable pig iron had now been secured but, remembering the difficulties which the first licensees had experienced, the steel manufacturers were not prepared to adopt the process. So, in 1858 Henry Bessemer with his partner, Mr Longsdon and brother-in-law Mr William Allen, together with Messrs Galloway of Manchester as equal partners, erected the Bessemer Steel Works in Sheffield.[3]

When they joined the partnership, Galloways surrendered the manufacturing licence which they had acquired in 1856 and this left the way clear for other companies to eventually adopt the converter process in the Manchester area. The Bolton Iron and Steel Company started steelmaking in the converter in 1863. This works installed 4×5 tonne converters in pairs in two pits and charged them from a bank of cupolas. With the addition of two Siemens open hearth furnaces in 1867 the plant became the largest steelworks in Lancashire. A close connection with steelmaking practice in the Sheffield area was soon established, for Mr Henry Sharp, who was General Manager and then Chairman of the company from 1889, was also General Manager of Samuel Fox and Co. Ltd., one of first licensees of the Bessemer process in South Yorkshire.[4] Even closer ties were established following his retirement, for in 1906 the company was taken over by Bessemer's own Sheffield business. The output of Bessemer steel was expanded under the new ownership and full production was maintained for over a decade. With the contraction of the market for steel in the aftermath of the first world war, however, the plant was unable to compete profitably with the larger works and was finally closed down in 1924.

The successful operation of the Bolton plant encouraged the formation of other companies to apply the converter process. The Mersey Steel and Iron Works at Toxteth in Liverpool had a long standing reputation for the production of wrought iron and, in 1853, became one of a small number of companies which made puddled steel. Bessemer steel was first made at the works in 1865 when two converters were built. Another four were installed in 1871. Ten converters had been installed by 1878,[5] all of 5 tonne capacity, but in the following year when the works were visited by the Iron and Steel Institute the Bessemer plant was only used intermittently and rail rolling had ceased. Four years later the plant was closed down. The cast iron church at Toxteth is the sole reminder of the once thriving iron and steel industry which was located there. Some of the Bessemer plant was bought by Alfred Hickman and re-erected at Bilston to form the nucleus of steelmaking at the Staffordshire Steel and Ingot Iron Works.

About 25 miles south of Manchester the London and North Western Railway Company had constructed a large railway manufacturing plant at Crewe in Cheshire. Four 5 tonne converters were initially erected here in 1867 but only two 3 tonne vessels were in use in 1878.[5] The Manchester Steel and Railway Works at Newton Heath installed four 3 tonne converters in 1871, but the plant had closed down by 1880.

The Lancashire Steel Company was in the process of building a small steel works with 2×5 tonne converters at Gorton near Manchester when the company became bankrupt. The Gorton plant was purchased and expanded by Bolckow Vaughan and Company, a large scale iron manufacturer located on the North East coast of Britain. The works became fully operational in 1872 and was used primarily to gain experience of the converter process. Additional converters were installed but the works did not make a profit and was closed down in 1874. Bolckow Vaughan transferred the Bessemer plant to the Eston works at Cleveland to form the nucleus for their steel production on the north east coast. The Manchester, Sheffield, and Lincolnshire Railway Company also operated two 3.5 tonne converters at Gorton for a period in the 1870s and 1880s.

In common with many of the early Bessemer steel works, a significant portion of the steel produced in Lancashire was fabricated into locomotive components, into tyres, wheels and axles for rolling stock and particularly into rails for the rapidly developing railway networks. A strong export market was developing for rails in the 1860s and the USA was a major customer. The works were fairly close to the Liverpool docks, which was the major terminal for shipping to America. But the goods had to be transported to the docks by railway until the Manchester ship canal was opened in 1893. The rail export trade had almost completely collapsed by that time.

CUMBERLAND SITES

Cumberland has a very rugged terrain and the iron works which had been built to exploit the local iron ore and coal deposits were all located on the narrow coastal strip. Small docks and harbours had been created at Barrow-in-Furness, Whitehaven, Harrington, Workington and Maryport for the transport of the local coal and iron ore in coastal ships of up to about 2 000 tonnes capacity. Same of the iron works which had been developed adjacent to the docks included iron rails in their list of manufactures. It was a logical development, therefore, for these works to adopt the converter process for the production of steel rails, using the local supplies of pig iron and coal, with coastal shipping to carry the goods for trans-shipment at a major port. Cumberland became a major growth area for the acid converter process.

The first converters in Cumberland were installed at Barrow-in-Furness. The Barrow Steel Company was formed in 1863 and the name was changed to the Barrow Hematite Steel Company in 1866, following the acquisition of the blast furnaces and iron ore mines of Messrs Schneider and Hannay. The first blast furnace had been erected at Barrow in 1859 and by 1863 ten blast furnaces were in operation, producing about 5 000 tonnes of pig iron per week and making this the largest iron works in the world, Fourteen furnaces were in blast by 1874 and the weekly output had increased to about 5 800 tonnes. The Duke of Devonshire, who became the first President of the Iron and Steel Institute in 1869, was Chairman and major stock holder of the company.

After small scale trials, a melting shop was built to house 5 tonne capacity converters in 1865. This was soon expanded and 6×5 tonne and 12×8 tonne converters were in operation a few years later. The plant was then described as the largest steel works in the world and 'The Sheffield of the North',[6] with an annual output of over 150 000 tonnes of steel. Two 12 tonne open hearth furnaces were built in 1880 to melt the in-house scrap produced in the rail mill. A heavy and a light plate mill were installed to make plates for the local ship building industry. Open hearth steel was generally preferred to Bessemer steel for this purpose, so the company developed a duplexing practice in which the blast furnace metal was partly blown in the converter to a medium carbon content and then transferred to an open hearth furnace for finish refining.

The output of steel was limited by the blast furnace production, which was restricted in turn by the rate at which iron ore could be extracted from the company's mines. Consequently, as the efficiency of the steelmaking operations improved and the annual output per converter was increased, the number of converters was reduced. Eleven converters were in use in 1879 and three of these were scrapped a few years later. When the melting shop was modernised in 1896 the remaining 8×8 tonne converters were replaced by 4×20 tonne vessels, which were mounted on a stage above the ground instead of in the conventional pit. These four converters produced a larger annual tonnage of steel than the original 18 vessels. The first inactive hot metal mixer to be built in Britain, with a capacity of 120 tonnes, was installed at the plant in 1890. The metal was still being cast into 2 and 5 tonne ingots and rolled into rails at the end of the century.[7, 8]

The adoption of the converter process about 50 miles further north along the Cumberland coast, in the main centre of hematite pig iron production, was surprisingly delayed until the 1870s.

The West Cumberland Hematite Iron Company at Workington started to build a steel works in 1870, but fell on hard times due to a temporary recession in rail orders. It was reformed as the West Cumberland Hematite Iron and Steel Company Ltd., in September 1872. Two months later the Bessemer shop started production with 4×7 tonne converters feeding a rail and plate mill. By 1878 2×5 tonne and 2×8 tonne vessels were in use.[5] George James Snellus, a graduate metallurgist who had worked at the Dowlais Iron and Steel Works and had recently toured American Bessemer plants, was recruited initially as manager of the melting shop, but was soon promoted to General Manager. Amongst the improvements to the established practice which he introduced was hot charging molten metal directly from the blast furnace into the converter, blending metal from two or three blast furnaces to obtain a suitable composition of the charge. This markedly reduced the costs by obviating the need for remelting the charge in a cupola. The improvements in efficiency were not sufficient, however, to counter the decrease in orders when the export market for rails declined rapidly in the late 1880s. The production from the plant decreased, with only intermittent operation, and it eventually closed completely in 1892.[9]

Members of the Cammell and Wilson families of Sheffield established a Bessemer works as a private venture in 1872 with converters, each producing 500 tonnes of steel per week at the village of Dronfield near Sheffield. The site proved to be too remote both from the supplies of hematite pig iron and from the docks when export orders for rail decreased and the profit per tonne dropped sharply. The situation was complicated commercially because the Cammell's and Wilson's were also major shareholders in the steelmakers, Charles Cammell and Co. Ltd., of Sheffield, a joint stock company. Cammell's Cyclops works in Sheffield and the Dronfield plant were major competitors and, to avoid questions being asked about divided loyalties, the Cammell and Wilson families sold the Dronfield plant to Charles Cammell and Co. Ltd., in 1882. The latter had also just purchased the Derwent Hematite Iron Company, which operated 3 blast furnaces at Workington. The entire works at Dronfield was dismantled and transferred, together with many of the employees, to the Workington site. Hot charging directly from the blast furnace, following the practice which Snellus had developed, was immediately adopted. In common with the West Cumberland and other works, however, the orders obtained for the rails fluctuated widely from year to year and there were times when the plant only operated intermittently. Trade improved in the early 1890s and two additional blast furnaces and two more converters were installed, but the profitability soon declined again. In 1909 the company was one of the main instigators in the formation of the Workington Iron and Steel Company Ltd., and the works was absorbed into the new company, which was known locally as 'the combine'. Steel making was terminated in the Derwent works and the melting shop was demolished.[10]

A 5 tonne converter was installed at Maryport in 1890 by the Hampton and Facer Special Steel Ingot Company Ltd. The intention was to manufacture carbon steels suitable for the tool and cutlery trades, in similar manner to Bessemer's original concept when he built his steel works in Sheffield. But the concept was found to be just as flawed here as it had been at Sheffield and the company had to sell the plant. It was acquired by Charles Cammell and Company Ltd. and eventually passed to the Workington Iron and Steel Company Ltd., but there is no evidence of further use of the converter.[11]

The Moss Bay Hematite Iron and Steel Company Ltd. in Workington is the site of the longest uninterrupted period of Bessemer converter operation in Britain. The company was formed in 1872. Four blast furnaces were in use when the Bessemer melting shop, with 2×7 tonne converters, and a rail rolling mill commenced operation in 1877. The decline in orders during the 1880s eventually resulted in bankruptcy and the appointment of a receiver in 1890. The company was more fortunate than the neighbouring West Cumberland works, however, for it was reformed with the same name under a new Board of Directors in the following year. The Bessemer shop was re-equipped with 3×8 tonne converters and a hot metal mixer was installed in 1901, but no other major changes to the melting facilities had been made before the company became a part of the newly created Workington Iron and Steel Company Ltd. in 1909.[11] The Moss Bay melting shop then became the only Bessemer plant in the new company.

The United Steel Companies Ltd. was formed in 1918 by the merger of two Sheffield steel makers and, in the following year, they acquired the Workington Iron and Steel Company Ltd. to secure their supply of hematite iron. The Bessemer plant continued to supply steel primarily for rolling into rails. For a period of time in the 1930s a daily train load of billets and slabs was despatched to the Bilston works of Stewart and Lloyds for the manufacture of pipes and tubes, but this trade ceased when the new Corby Bessemer plant was fully operational.

The original 8 tonne converters had been replaced by three of 15 tonne capacity. These in turn were replaced by 2×25 tonne converters and a 400 tonne hot metal mixer was installed when a new melting shop was opened in 1934. This plant continued in operation when the industry was nationalised in 1969 and the United Steel Companies became a part of the British Steel Corporation. The melting shop was eventually closed in 1974, thus ending 97 years of acid Bessemer steel making at the Moss Bay works and over 100 years of production in the Workington area.[12] Throughout the post war period, from 1945 to when the melting shop was closed, the works had produced an average of approximately 250 000 ingot tonnes of steel per annum by the acid process. This was less than half the output which had been achieved in the peak production years at the start of the 20th century but, after the wholesale closure of the other British Bessemer plants in the 1920s and 1930s, Workington was the only acid Bessemer steel produced in Britain.[13] One of the converters is now permanently displayed at the entrance to the Kelham Island Industrial Museum in Sheffield.

ACKNOWLEDGEMENTS

The author is grateful to the many people and particularly to Mr Trevor Lodge and Mr Arthur Little who have supplied information which is incorporated in this chapter.

REFERENCES

1. *Sir Henry Bessemer, An autobiography*, reprinted 1989, The Institute of Metals, 176.
2. H. Bessemer: *ibid*, 175.
3. H. Bessemer: *ibid*, 179.
4. Portraits and pen sketches of leading men, One hundred men of Lancashire, Biographical Publishing Co., London, 1900.
5. J.S. Jeans: *Steel, its history, manufacture and use*, E. and F.N. Spon, London, 1880.
6. C,McCombe: *Foundry Trade Journal*, June 21, 1984, 563.
7. J.M. While: *J. Iron Steel Inst.*, 1901, **1**, 299–305.
8. Anon: *Iron Coal Trades Rev.*, August 4, 1899, 205–210.
9. J.Y. Lancaster and D. R. Wattleworth: *The iron and steel industry of West Cumberland*, British Steel Corporation, Teeside Division, 1977, 47–52.
10. J.Y. Lancaster and D. R. Wattleworth: *ibid*, 81–92.
11. J.Y. Lancaster and D. R. Wattleworth: *ibid*, 102–103.
12. J.Y. Lancaster and D. R. Wattleworth: *ibid*, 66–80.
13. J.Y. Lancaster and D. R. Wattleworth: *ibid*, 164,

The Bessemer Process in South Wales

M.E. WALKER AND R.D. WALKER

INTRODUCTION

For just over a century, from 1865 to 1969, the Bessemer process and processes derived from it were used in eight steelplants in South Wales (see Fig. 1).

The acid Bessemer was most commonly employed in South Wales, for although the early work that led to the evolution of the basic Bessemer or Thomas process took place at Blaenavon, it was used sparingly in its country of origin. In its final fling in the Welsh steel industry, the converter process employed in the Abbey Works of the Steel Company of Wales incorporated yet another radical departure from classic practice by relying on oxygen/steam blast in the VLN (very low nitrogen) plant.

In what follows, the experimental period up to 1865 is first described and this is followed by a plant-by-plant account of over 100 years of pneumatic bottom-blown steelmaking in South Wales. The locations of the plants are shown in Fig. 2 and the dates in the sub-titles refer to periods when bottom-blowing was practised.

THE PIONEERING PERIOD BEFORE 1865[2–7]

Until the late 1860s the prosperity of British ironworks was based increasingly on the manufacture of wrought iron rails. For example, it was said that between 1850 and 1861 the company at Ebbw Vale made a profit of almost a million pounds. There was therefore a great deal of money to be made, but since wrought iron as an engineering material has its limitations, the more far-sighted were already looking to the future. Two companies, one at Dowlais near Merthyr and the other at Ebbw Vale, carried out experiments in steelmaking before the Bessemer process was established on a production basis in 1865.

At Dowlais, the ironworks developed by the Guests had fallen into disrepair at the end of the 1840s. Throughout most of that decade the company existed under the threat of financial ruin at the hand of its leaseholder, the second Marquis of Bute, for its lease expired on 1 May 1848. Lord Bute wanted to expand the port of Cardiff on a grand scale, financed partly by what he considered was justifiable compensation for a century of low royalties. Negotiations began in the early 1840s, but the Guests were unwilling to accept terms which they considered unreasonable, and prepared to leave Dowlais. By February 1848 three of the blast furnaces had been blown out, and rails and mechanical equipment were brought up from the mines to be sold, but some three weeks later Lord Bute died unexpectedly.

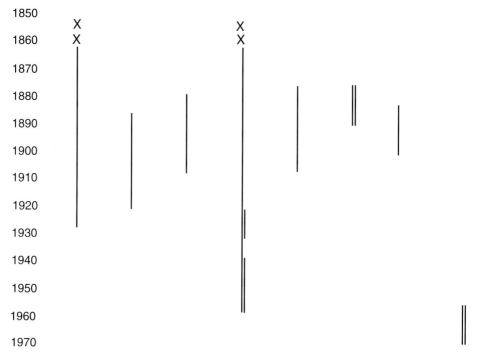

Fig. 1   Chronology vs time line of Bessemer converters in South Wales steelplants
(X = experiments; single line = acid; double line = basic).

Thus the Dowlais Iron Co. was saved, but Sir John Guest's death in Dowlais in November 1852 meant that the new lease had to be signed by his wife, Lady Charlotte, in January 1853. The restoration and improvement of the works that followed allowed the Dowlais Iron Company to remain the leading ironworks in Merthyr; indeed its capacity was almost equal to that of its Merthyr rivals combined. Nevertheless a senior manager, possibly the newly promoted William Menelaus, concluded in June 1856 'that the works can only be restored to a fair condition by a very large expenditure.'

It was clear that the senior management at Dowlais was ready to consider innovation, both to retain the position of the company as a leading ironmaker and to maintain its profitability. The importance of Bessemer's Cheltenham paper was immediately noted by Lady Charlotte, and by Menelaus. With the assistance of Edward Riley, the works chemist, and his own assistant Edward Williams, Menelaus oversaw the production of two 25 foot bars of steel and, as a result, the trustees of Dowlais met with Bessemer on 27 August, 1856 and secured the first licence to use the Bessemer patent. For £10 000 they were allowed to produce 20 000 tons per year for the next ten years, with an extra 1/4d per ton up to 70 000 tons.

Trials were then carried out at Dowlais. A converter was constructed and placed opposite a furnace making iron for puddling to rails. The molten iron was run into the converter, and

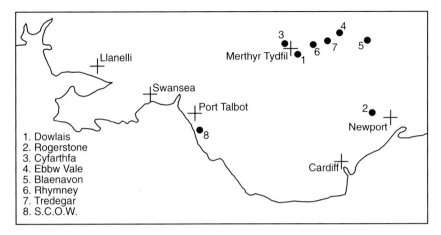

Fig. 2   Bottom blown steelplants in South Wales (adapted from Ref. 1).

hot air blasted through it. After the expected violent reaction the metal was successfully rolled into a small bar which was nevertheless found to be very brittle. In 1858 the first steel rail was rolled in Dowlais, but broke while still hot, much to the delight of the puddlers.

The brittleness of this early steel was caused by the presence of phosphorus originating in the iron ores smelted in the blast furnaces. Bessemer had not realised that the process would not work with pig iron containing a high level of phosphorus, that is, above 0.06%. Riley's analysis of the early failures show a phosphorus content as high as 0.753% in the steel produced. It took several years for the phosphorus problem to be recognised and the need for suitable non-phosphoric ores to be established. Eventually on 5 June 1865, the first ingots of Bessemer steel produced at Dowlais were cast, rolled into rails on 10 June and tested by the London and North West Railway in the following September. The pioneering phase was over.

Earlier, at the Ebbw Vale Company, during 1854 and 1855, the works chemist George Parry experimented by blowing air through pipes fixed in the bottom of a reverberatory furnace holding molten iron. This process was invented by J G Martien of New Jersey, but the patent rights were bought by Messrs Darby and Company (owners of the Ebbw Vale Company) in 1855. Parry succeeded in making steel, but the process was abandoned because the wear on the furnace made it impossible to contain the molten metal. Not surprisingly, however, the Ebbw Vale managing partner, Thomas Brown, realised the importance of the Bessemer process, and in 1856 he immediately went to see Bessemer, with the hope of gaining rights to use the new process. He wanted to buy the exclusive rights for £50 000, but Bessemer refused, preferring to issue licences to one major operator (including Dowlais) in each region.

Undeterred, Brown sought the help of Robert Mushet at Coleford and at Mushet's suggestion built a small Bessemer converter at Ebbw Vale. Problems arose, however, in trying to replicate Bessemer's achievement, and the ingots they produced cracked on forging. Mushet recognised the need for deoxidation and added spiegeleisen to a re-melted ingot at Coleford before returning it to Ebbw Vale where it was rolled into a perfectly sound, double headed steel rail. Mushet took out a patent on his deoxidation process in 1857, but possibly on the advice of Thomas Brown he substituted the name of Martien for Bessemer. Subsequently Mushet refused to assist Bessemer out of loyalty the Ebbw Vale Company and relations between the two men deteriorated. In 1860, however, the two trustees of Mushet's patent forgot to pay the third year's stamp duty of £50 and consequently the process became public property and Bessemer had the right to use it.

Also in 1860, George Parry used an improved regenerative furnace in combination with a re-melting furnace equipped with tuyeres to make steel from wrought iron and scrap. First, he melted and re-carbonised the charge with coke and then blew air through the tuyeres to burn the carbon out. This process was patented in 1862 and Ebbw Vale planned to use it in conjunction with Mushet's process in a way that threatened Bessemer's position as the inventor of pneumatic steelmaking. Bessemer could probably have won the argument in the courts, but at considerable cost of time and money. Instead, in 1864 he paid the Ebbw Vale Company £30 000 for both the Parry and Martien patents. Parry received £10 000 and the Company abandoned its plans, preferring instead to construct a Bessemer plant consisting of six 10 ton converters.

DOWLAIS (1865–1930)[8–9]

Recognising the very poor return Dowlais had made on its original outlay, Bessemer in 1865 consented to the negotiation of a new licence, agreeing to forego £20 000 from the first royalties. Meanwhile Menelaus embraced the new technology with caution, gradually increasing the manufacture of steel from 2 257 tons in 1866 to 21 179 in1870 and of rails from 762 tons in 1866 to 16 967 in 1870. By then six 5 ton converters had been installed and in 1868 Menelaus argued that 'even the best wrought iron ever made was far inferior to the Bessemer metal, at least for rails.' It was true that steel rails were, originally, more expensive to produce, but the expiry of the Bessemer patent in 1870 meant that by the early 1870s the differential between steel and iron rails had narrowed to less than £4 a ton.

Further developments at Dowlais included the erection of Siemens-Martin open hearth furnaces, which extended the range of steel products. For rails, however, the Bessemer process was unrivalled, and when Menelaus died in 1882 the Bessemer converters had produced 113 433 tons of steel ingots. Indeed after 1883 no more iron rails were made at Dowlais and three mills were laid down for the manufacture of tin plate. By 1883, however, these were abandoned in order to gear the whole production to steel rails: a new mill constructed in the 1880s was capable of turning out 5000 tons of steel rails a week. In 1885, a plant was laid down for the manufacture of steel sleepers which had the merit of being resistant to tropical termites and thus they rapidly became popular in the colonies.

Meanwhile, the management of the company was actively considering reducing the costs inevitably incurred by an inland site, and relocating to the coast near Cardiff. Such a move would, moreover, open up the possibility of installing the latest machinery and processes. Yet the plan, admirable in its conception, would certainly have had the unfortunate consequence of causing widespread unemployment and consequent hardship to the population of Dowlais, and this was the sticking point as far as Lord Wimborne, eldest son of Sir John

Guest, was concerned. Some kind of employment had to be maintained for the people of his native town.

Thus, while first the new ironworks and then the steelworks were established at East Moors, a greenfield site next to the thriving Cardiff docks and with easy access to the main railway line to Paddington, rail production remained at Dowlais. By 1895 this consisted of two Bessemer steelworks: the original plant, which had steadily been updated and which in 1895 had four 10 ton converters, and the new works with two 15 ton converters. The new rail mill, opened in that year, was capable of producing 5000 tons of steel rails a week, with fewer men than had been required to make 600 tons of iron rails.

Meanwhile significant changes were taking place in the nature of the company. Lord Wimborne, now more attached to his Dorset estate than to the responsibilities involved in the running of the company, eventually sold out to Arthur Keen, who joined with Nettlefolds in 1902 to form one of the great engineering companies of the twentieth century.

As far as Dowlais was concerned great efforts were made to keep the enterprise alive, with a new rail mill opening in 1905 and a morale-boosting visit from the King and Queen in 1911. The Great War brought prosperity, but at its end the anomalous position of the steelworks could not be disguised. The slump of 1921–22 involved a fall in the price of rails from £25 per ton in January 1921 to £8 15s in October 1922. There was a slight revival up to 1929, but the need to convert to electricity from steam, and the gradual realisation that the decline of the staple industries of the nineteenth century was irreversible, brought little hope to the steelmakers. UK orders for rails dried up, and an unsatisfactory consignment of rails to Egypt sealed the fate of the company. In 1929 a company report doubted the viability of either steelworks, so that an approach from the rival steelmakers Baldwins was greeted with alacrity.

This was a death knell for the Dowlais works. The newly-formed company, Guest, Keen and Baldwins lacked the nostalgic links which had saved Dowlais at the end of the nineteenth century and despite efforts to get orders, the new enterprise decided in 1930 that Dowlais should close. For years the works lay derelict, as a kind of awful memorial to the people who had worked there.

CASTLE WORKS, ROGERSTONE (1888–1921)[8, 9]

The creation of Guest, Keen and Nettlefolds (GKN) in 1902 involved the absorption of the Bessemer plant at Rogerstone so it seems logical to deal with it now because of its links to Dowlais.

Nettlefolds was a Midlands company which dominated the woodscrew market and in the second half of the nineteenth century it had been experimenting with the use of steel, rather than wrought iron. To replace wrought iron was potentially costly, but it was realised that the superior qualities of steel would make such a transformation inevitable. The company's small steelworks at Hadley, in Shropshire, was unlikely to provide an adequate supply of steel and what was needed was a new larger works, close to a supply of pig iron and with good communications.

In 1885, Edward Steer, one of the Nettlefolds directors, was dispatched to South Wales, where he soon found an ideal site at Rogerstone in the Ebbw Valley. Here there was a tradition of ironmaking, an excellent water supply, rail connections with the valley collieries and the Midlands, and access to Newport docks. In 1886 Steer moved to Rogerstone so that he could oversee the building of the new works, which was to include an acid Bessemer plant charged with pig iron purchased from various South Wales works. The move was so successful that the Nail Works was moved from Birmingham in 1905, while wiremaking was transferred to Newport in 1901, thus basing all Nettlefolds' 'heavy end' of its production processes around its Rogerstone steel works.

Arthur Keen had long shown interest in the thriving Nettlefolds concern, partly in order to integrate the manufacture of steel goods. However after the creation of GKN, it was discovered that Dowlais steel was unsuitable for wire-drawing, so that high quality ore had to be purchased from Westmorland to meet Nettlefolds' requirements. Nevertheless the Rogerstone business continued, the two Bessemer converters lighting up the streets at night. It was the First World War that was the immediate cause of the closure of steelmaking at Rogerstone. The Bessemer process was to prove too costly when scrap from the battlefields of Europe provided cheap raw material for the open hearth furnaces. In March 1921 the last steel was made and the billet mills closed later.

CYFARTHFA (1885–1910)[8]

Cyfarthfa too eventually came within ambit of GKN. It had once been the greatest ironworks in the world, but by 1880 it was clear that unless a steelworks was added its future was uncertain. Therefore four 10 ton Bessemer converters were installed, arranged in line, not under cover, with the blast provided by two blowing engines. One such converter is shown in Fig. 3.

However the operating costs were such that eventually the undertaking had to seek incorporation as a limited liability company. Moreover, the position of the works was no better than that of Dowlais, so that not long after the formation of Guest, Keen & Co., Arthur Keen started to negotiate with the company.

His reasons were twofold. In the first place he could gain access to their collieries, but secondly the existence of the Cyfarthfa steel operation, smaller and less efficient than those of either Dowlais or Cardiff, would to some extent shield them should cutbacks be necessary. In 1903 it was decided that the works should continue 'for a time' but matters did not improve and the works closed in 1910.

EBBW VALE (1865–1929)[6, 10–12]

Erection of the steelworks in the period 1866–68 was supervised by E. Windsor Davies who became general manager on the retirement in 1873 of Abraham Darby, the managing director. The remaining Ebbw Vale directors, however, had little interest in steel and plant performance thereafter was poor, with little change in practice over a period of more than 30 years. Eventually, in 1905–6, the Bessemer plant of 1868 was demolished and replaced by a

Fig. 3   Ten ton capacity converter during a blow at Cyfarthfa around the turn of the century. (Photograph supplied by the Welsh Industrial and Maritime Museum, Cardiff: acquisition number 76.751/60)

German-built installation consisting of two 20 ton converters and two mixer furnaces, one of 250 tons capacity and the other, the biggest in the world at that time, of 750 tons capacity. Weekly steel production from the Bessemer plant was 4000 tons with a further 1000 tons coming from five open hearth furnaces. Ebbw Vale flourished for a time, but its inland site, 20 miles from the seaport of Newport was a considerable handicap adding extra transport costs to both iron ore imports and to steel exports. The commitment of the Ebbw Vale Company's directors to steel was lukewarm, since they could make greater profits from their collieries and in April 1911, at the instigation of Sir D. A. Thomas, the blast furnaces and steelworks were shut down. The expected profits from coal did not materialise, however, and the iron and steel plant re-opened in April 1912. It was then virtually in full production up to and throughout the First World War when amongst other products, the output of heavy galvanised sheet rose from 600 to 1000 tons per week.

A post-war boom in steel demand lasted until the end of 1920, but the following decade was a troublesome period in the history of Ebbw Vale. Lack of demand for steel, industrial unrest and colliery disasters all contributed to uncertainty, but late in 1927 the steelworks and collieries were again momentarily busy. Limited modernisation, including the installation of a 1500 ton mixer furnace in the Bessemer plant was begun. Alas, this optimism proved to be unfounded and the steelworks closed in October 1929. Nevertheless, the sheet mill at Ebbw Vale continued in operation, rolling bought-in steel and in 1933, there was a change that would prove to be highly significant for the South Wales steel industry when the mill began to produce sheet steel for motor car bodies.

BLAENAVON (1881–1909)[7, 13, 14]

In 1870, the students attending a course of lectures at the Birkbeck Institution delivered by Mr George Chaloner heard him say that 'the man who succeeds in eliminating phosphorus in the Bessemer converter would one day make his fortune.' Amongst those present was Sidney Gilchrist Thomas who took this remark very much to heart. Thomas was a very bright 20 year old who, on the death of his father three years earlier, had abandoned plans to pursue a medical career and instead became a clerk in the Metropolitan Police Courts in order to supplement the family income. He soon mastered the work, but his real interest was science and so he moved to the Thames Police Court where he made an arrangement with a colleague that gave him two days a week free. This enabled him to establish his credentials as a chemist by taking and passing every examination he could at the School of Mines. In 1873, he began to make regular contributions to a journal called '*Iron*' and in 1874 during his annual holidays he went to South Wales, staying with his cousin, Percy Carlyle Gilchrist, then a chemist at Cwmafan Works, Glamorganshire. Thomas visited the works at Landore near Swansea where Siemens had installed his open-hearth process as well as other plants. He returned to South Wales in 1875 and soon after started a series of experiments with Chaloner at the laboratories of the Birkbeck Institution. The subject of Thomas' research was the durability of various furnace lining materials and he was particularly interested in lime, magnesia and magnesian limestone. He conceived the idea that a basic lining material would not only enable the phosphorus to be separated from the iron during steelmaking, but would also make it possible to retain the phosphorus in the slag which could then be used as an agricultural fertiliser.

Thomas conducted some experiments at his home which he was at grave risk of setting alight and it was obvious that he needed a more suitable location for his investigations. He applied for a vacant chemist's post at Blaenavon Works, but was considered too inexperienced. Fortunately for Thomas, Percy Carlisle Gilchrist, his cousin, was appointed and Thomas immediately sought to enlist his help in making some tests. Initially, Gilchrist was reluctant to do much, until in 1877 he too began to share Thomas' enthusiasm and made many experimental 'blows' in a miniature converter in their mountainside shed. Thomas communicated with his cousin by letter and increasingly he made flying week-end visits to Blaenavon, often leaving London on the midnight train on Thursday and returning in time for the Court on the following Monday morning. The technical level of Thomas' letters is illustrated by a query on 17 September, 1877 about the temperature at which 'SiO_2 begins to displace P_2O_5 from Phosphate of iron.' Their first patent was taken out in November 1877. Almost inevitably, the secrecy with which the two inventors conducted their affairs was breached and early in 1878 the manager of the Blaenavon Company, Mr E. P. Martin demanded an explanation of their clandestine activities. He was favourably impressed with their work and even agreed to purchase a share of the patent. Through Martin's intervention, larger-scale experiments took place at both Blaenavon and at Dowlais, where Menelaus made a converter available. Thomas continued to patent every aspect of his discoveries and in March 1878 he made his first public pronouncement about his new process. This took place at the Spring meeting of the Iron and Steel Institute at which Mr I. Lowthian Bell read a paper entitled 'The Separation of phosphorus from Pig Iron.' Various worthies contributed to a discussion of dephosphorisation and it was not until near the end that Thomas got a chance to speak. His claim to 'have been enabled, by the assistance of Mr Martin at Blaenavon, to remove phosphorus entirely by the Bessemer converter was received with some considerable reserve. Undeterred, Thomas wrote a paper '*On the Elimination of Phosphorus in the Bessemer Converter*' for the Autumn meeting of the Iron and Steel Institute in Paris. No time could be found for Thomas to read his paper, however, and it was not until May 1879 that it was discussed. Its key paragraph is as follows.

> '*After a very extended series of trials it was found that by firing bricks made of aluminosiliceous limestone at a very intense white heat, a hard and compact basic brick is formed. These bricks, unfortunately, labour under the defect of a liability to disintegration when exposed to the action of steam. By the use of certain aluminous magnesian limestones – and equivalent combinations – this difficulty has been, after many failures, overcome. The problem is solved by substituting a reasonably durable basic lining for the former siliceous, and therefore acid, one and by avoiding waste of lining by making large basic additions so as to make a highly basic slag at an early stage of the blow.*'

The last sentence is a concise description of the principle of the Basic Bessemer or Thomas process.

Not everyone was dismissive of Thomas. E. Windsor Richards, manager of Bolckow, Vaughan and Company in Middlesborough, read the paper offered to the Paris meeting in 1878 and discussed its content with Thomas during a visit to the works at Le Creusot. The upshot of this encounter was that Mr Richards and Dr Stead, Consulting Metallurgist to Bolckow, Vaughan and Company visited Blaenavon in October 1878 and saw for themselves that Thomas' claims were genuine. After discussing the experimental work with Menelaus at Dowlais, Windsor Richards returned to Middlesborough and convinced the directors of the company that they should provide appropriate facilities for Thomas and Gilchrist who with the agreement of Mr Martin thereafter continued their activities in the North East of England. The link with Blaenavon was therefore broken, but the site where the developments that led to large quantities of European iron ores becoming available for steelmaking is marked by a memorial erected in 1960 through the efforts of members of the Newport and District Metallurgical Society.

Paradoxically, although the Basic Bessemer or Thomas process has its origins in Blaenavon Works, which became the Blaenavon Company Limited in 1879, steelmaking at this works relied on the acid Bessemer process and later on from around 1909 on the open hearth. The iron ores used were either local hematite ores or they were imported from Spain and the analysis of Blaenavon pig iron varied in the range 3.1–3.6%C, 1.4–2.5%Si, 0.02–0.07%S, 0.45–0.46%P and 0.8–1.3%Mn, typical of acid Bessemer practice. The 20th century was difficult for the Blaenavon Company Limited and steelmaking ceased in 1922, although from time to time the mills rolled bought-in ingots. Open hearth steelmaking restarted in 1937 and steel was produced until 1948 when the works finally closed.

RHYMNEY (1878–1891)[15] AND TREDEGAR (1882–1901)[16]

The roots of both these works lay in the iron trade, but the decline in demand for wrought iron rails and its replacement by steel necessitated change. Accordingly, the Rhymney Iron Company converted its ironworks to steelmaking in 1878 by installing three 7 ton Bessemer converters. A notable feature of this plant was that it used the American system of placing the converters side by side instead of facing each other, an arrangement that was said to facilitate fettling since the sparks from the converter in blast did not inconvenience workers repairing the other vessel. The planned capacity was 500 tons per week, but sometimes double that output was achieved. Two further converters were installed in 1879 to work the basic process. Even so, the cost of operating an inland site proved too much and the works closed in 1891.

   The Tredegar Iron and Coal Company Ltd was formed in 1873 with an impressive list of directors including William Menelaus and Sir Lowthian Bell. Their motivation was similar to that of the Rhymney Iron Company and two 8 ton Bessemer converters were installed and began production in 1882. Tredegar steel acquired a reputation for high quality, but a combination of competition from plants using the open hearth process and the high transport costs of an inland site made Tredegar uncompetitive. The works closed in 1895, re-opening for re-rolling in 1896 and staggering on until it was finally closed in 1901. The acquisition of the Tredegar Iron and Coal Company's facilities by the brothers L D and A V Whitehead links that company to the present day through the use of the familiar name Whiteheads to refer to the British Steel re-rolling plant in Newport.

EBBW VALE (1938–1962)[6, 17–20]

The Ebbw Vale Steel, Iron and Coal Company closed down in 1929, having failed to survive the recession, but later when Richard Thomas and Co., acquired the plant only two blast furnaces were thought to be worth saving, the rest of the plant being demolished. By 1935, Richard Thomas and Co., owned about half of Britain's tinplate production capacity and consequently they decided to follow US practice by building a hot strip mill, since a continuous mill of this type was much faster and more economical than the traditional hand mill method. Initially, the company wanted to build the new mill at its Redbourne Iron & Steel Works in Lincolnshire, but political pressure forced them to site it at Ebbw Vale where unemployment was high.

   The works was completely rebuilt in the period 1936–38. New coke ovens and by-product plant were constructed and on the basis of its success at the Corby Works of Stewarts and Lloyds Ltd, the basic Bessemer or Thomas process, as it is perhaps more fittingly referred to in connection with Wales, was installed at Ebbw Vale along with three 75 ton open hearth furnaces and an inactive mixer. There were three 25 ton converters and, although the original intention was to use them only to desiliconise hot metal for the open hearth shop, refined metal was in fact manufactured from 1938 onwards. Later, two more open hearths were added and the mixer capacity was increased. Iron making practice at Ebbw Vale had the aim of producing low-sulphur hot metal (0.04–0.08%S). As a consequence, the blast furnaces were operated at a high hearth temperature and with a basic slag, conditions which are well known to lead to high silicon contents in the iron. A typical iron analysis was 3.6%C, 0.85%Si, 0.06%S, 1.55%P and 0.8%Mn. This phosphorus content is a little lower than was usual in Thomas practice and although the extra silicon was some compensation, skulling was sometimes a problem.

Around 90% of the converter shop's output was extra soft rimming steels, the remainder being silicon steels and steels recarburised to 0.08 to 0.10%C. A notable characteristic of Ebbw Vale steels was their low nitrogen contents (low relative to other Thomas plants, that is, for it has to be recognised that open hearth steel was superior to Thomas steel in respect of nitrogen). Two factors contributed to this situation. First, the use of mixers meant that the temperature of the iron delivered to the converters was low, around 1180°C, and secondly, the two rotary blowers could not work in parallel so that the flowrate of the air blast was low, prolonging the blow to over 20 minutes. One third of the casts had <0.008%N, half had <0.012%N and only 8% were downgraded on the basis of their nitrogen content. Blowing capacity was also the main factor limiting production. With one converter in operation, 15 charges per shift (1000 ton a day and 5 200 ton a week) were possible. With two converters in service, the impossibility of blowing both at the same time held production at between 22 and 26 charges per shift with knock-on effects on daily and weekly output.

The period following the Second World War saw intensive developments in the use of oxygen in steelmaking in various parts of the world. In the 1950s, oxygen enrichment of the blast was practised at Ebbw Vale to lower the nitrogen content of the steel. Some form of temperature control was necessary, however, to avoid excessive lining wear in the lower half of the converters and so in 1958 the use of oxygen/steam mixtures was introduced. This development was quickly over-shadowed, however, when one of the bottom blown converters was replaced by a 35 ton top blown converter in 1960 and Britain's first LDAC steelmaking process began at Ebbw Vale. Within two years, in March 1962, a second Thomas converter was given a solid tar dolomite bottom and an overhead oxygen lance and operated this time as an LD vessel without lime blowing facilities. Later in September 1962, a third Thomas converter was modified and operated as an LDAC vessel and the link with Bessemer was broken. Steelmaking at Ebbw Vale ceased altogether in 1978, although the works remains in operation as an arm of British Steel Tinplate.

STEEL COMPANY OF WALES (1959–69)[21, 22]

Several factors led the Steel Company of Wales to choose the VLN process to provide the extra 12 000 ton of steel per week required by the new universal mill installed in its Abbey Works at Port Talbot in 1959. The need was for high quality steel suitable for deep drawing applications and the ready availability of tonnage oxygen pointed to some kind of pneumatic process, especially since this type of plant not only has relatively low capital costs, but low conversion costs as well. High phosphorus iron ores were in plentiful supply and the existence of open hearth furnaces on the same site to consume scrap meant that operating a bottom blown process was feasible. Although the VLN had been employed successfully by CNRM in Belgium and by Klockner AG in Germany, the plant installed at the Abbey Works was unique in being the only plant that had no air blower, but instead relied exclusively on a blast of oxygen and steam.

Three 60 ton converters were installed and operations began at the end of June in 1959. Although various combinations of rammed dolomite and magnesite bonded with tar were used to line the converters, best results were obtained with tar bonded dolomite on its own. The average composition of the hot metal was 3.8%C, 0.35%Si, 0.024%S, 1.8%P, 0.8%Mn and 0.005%N, the high phosphorus content necessitating a two-slag process. The first blow, which lasted 10 minutes, was carried out using an oxygen/steam ratio of 1.5:1 dropping to 0.7:1 during dephosphorisation. At this stage, the metal composition was around 0.025%C and 0.04–0.07%P. After about two thirds of the slag had been run off, more lime was added and the charge blown again to obtain with the average composition of 0.05%C, 0.022%S, 0.022%P, 0.30%Mn, 0.0011%N, 0.03%Cu, 0,05%Ni and 0.01%Sn. The average charge to tap time was 30 minutes and one converter could blow 17 heats in an 8 hour shift. The nitrogen content of the steel was consistently very low with 15% of heats being below 0.008% and steels made by the VLN process were considered to be equivalent to open hearth steels for all applications.

One disagreeable feature of the VLN process was its propensity to discharge clouds of iron oxide particles which settled as a red dust on every available surface in the vicinity of the works and was reputed to have spoiled many a line of washing when the wind blew inland towards Port Talbot.

In 1969, two 300 ton LD converters were commissioned in what was now the Port Talbot Works of the British Steel Corporation, coinciding with the closure of the VLN plant and signalling the end in South Wales of the bottom blowing tradition begun at Dowlais and Ebbw Vale over a century earlier.

ACKNOWLEDGEMENT

This chapter could not have been written without the support of Robert Protheroe Jones of the Welsh Industrial and Maritime Museum. He generously made the Museum's archives available to the authors and was a ready source of advice and guidance. Nevertheless the final text is the authors' and one of us (RDW) takes full responsibility for it.

REFERENCES

1.   R. Protheroe Jones: *Welsh Steel,* National Museum of Wales, 1995.
2.   Mandy Walker: *From Dowlais to Tremorfa,* Tremorfa Books, 1993.
      (available through the National Museum of Wales).
3.   John A. Owen: *History of Dowlais Works 1759–1970,* Starling Press, 1977.
4.   M. Elsas: *Iron and Steel in the Making, Dowlais Iron Company Letters,* 1782:1860,

Glamorgan Record Office, 1960.

5. John A. Owen: *History of Dowlais Works 1759–1970,* Starling Press, 1977.
6. A. Gray-Jones: *A History of Ebbw Vale,* 1970,
    (made available by the Welsh Maritime and Industrial Museum).
7. K. C. Barraclough: *Steelmaking 1850–1900,* The Institute of Materials, 1990.
8. Edgar Jones: *A History of GKN vol 1 1759–1918,* Macmillan, 1987.
9. Edgar Jones: *A History of GKN vol 2 1918–1945,* Macmillan, 1987.
10. *The Ebbw Vale Works Magazine*, 1922, **2**(5). and 1923, **2**(8).
11. F. J. Ball: *Notes on the History of Ebbw Vale Iron works (1843–68),*1987,
    (made available by the Welsh Maritime and Industrial Museum).
12. Keith Thomas: *Ebbw Vale in Old Photographs,* 7 volumes, 1979–96, Kenn Publishers, Ebbw Vale.
13. Tom Grey-Davies: *Blaenavon and Sydney Gilchrist Thomas,* The Historical Metallurgy Society, 1978.
14. F W Harbord: 'The Thomas-Gilchrist Basic Process 1879-1937', *JISI*, 1937, **136** (2).
15. R Laybourne, *The Bessemer Steel Plant at Rhymney,* Transactions of the South Wales Institute of Engineers, Session 1877–78.
16. W Scandrett: *Old Tredegar vol* 1,1990, Starling Press.
17. Comprehensive report of the B.I.S.R.A. Sub-Committee, *Bessemer Steelmaking,* Iron and Steel, July 1949.
18. M Demarteau: *The Ebbw Vale Works,* 1958,
    (made available by the Welsh Maritime and Industrial Museum).
19. C. Bodsworth: 'Richard Thomas and Baldwins Ltd', *Ironmaking and Steelmaking*, 1996 **23** (6).
20. D. J. Jones, A. E. Parsons and N. Morris: 'LD and LDAC Operating Experience in Britain', *Journal of Metals,* August 1963.
21. Anon.: 'S.C.O.W. Extensions', *Iron and Steel,* May 1956.
22. M. C. Harrison and P. Truscott: 'Operation of the oxygen/steam blown converters at the VLN steel plant of The Steel Company of Wales Ltd', *JISI*, August 1961.

> This is dedicated to the memory of the late
> Mandy Walker
> who carried out the research for this chapter and
> prepared most of the text, but sadly was unable to
> complete the work due to her untimely death on
> 12 December 1997

Basic Bessemer Steelmaking: Cleveland and Beyond

T. J. LODGE

SYNOPSIS

Unlike the vast majority of early UK acid Bessemer plants, which supplied to the burgeoning railway industry, the later basic Bessemer works were not united by a similar common focus. If there was a commonality, it was a desire of existing iron producers to become associated with all conquering steel.

Consequently, this chapter has been prepared in two distinct parts. The first details the pioneering work done by Thomas and Gilchrist, together with the commercial introduction of the basic process on Teesside. The second charts chronologically the spread of the basic Bessemer process through the UK – from Teesside to Sheffield, the West Midlands, the Clyde Basin, West Yorkshire, and ultimately, in the present century, to the East Midlands. Basic Bessemer developments in South Wales are dealt with in that chapter.

DEVELOPMENT AND COMMERCIAL INTRODUCTION

For some twenty years phosphorus (and, to a lesser extent, sulphur) remained the *bête noire* of the Bessemer process. Indeed, it put such a strait-jacket on the process that it dictated the very geography and morphology of the UK's developing steel industry. The eventual solution by cousins Thomas and Gilchrist – the use of a basic furnace lining and slag – is almost too well known to need repeating here. However, a brief description is necessary since the development was the key to the phenomenal expansion of bulk steelmaking from the 1880s onwards – not so much by the Bessemer process but increasingly so by the Siemens open hearth melting process. A fuller account of this phase can be found in the works of Barraclough and Hempstead, cited in the bibliography at the end of this chapter.

Sidney Gilchrist Thomas was a police court clerk in London who studied chemistry and related subjects at evening classes. He was inspired to find a solution to the phosphorus problem that plagued the Bessemer steel industry by George Chaloner, his chemistry lecturer at the Birkbeck Institute. Chaloner had said 'The man who eliminated phosphorus in the Bessemer converter could one day make his fortune.' When Thomas began his experiments in 1875, his cousin Percy Carlyle Gilchrist was already working as an analytical chemist in South Wales. Following his move to Blaenavon Works, Gilchrist undertook experiments in his spare time under Thomas' directions. The experiments were supported by the Works Manager, Edward Pritchard Martin, who allowed the erection of equipment in which 150–200 kgs of pig iron could be blown down. Thomas' first patent – the use of sodium silicate as a binder for the chalk, lime or magnesia lining – was taken out in November 1877. William Menelaus was alerted to the potential of the new process and he arranged for trials at Dowlais Works in a 7 ton commercial Bessemer converter fitted with a basic lining. By September 1878 the cousins had found that burnt lime gave superior results, and they patented this as a feature of manufacturing the lining on 7th October 1878.

In the meantime, Thomas had announced his success in removing phosphorus to the Iron and Steel Institute's London meeting in March 1878. Apparently his words fell on deaf ears – most of those present must have thought the youthful Thomas something of a crank, and hardly likely to have succeeded in solving a problem which had baffled the country's leading metallurgists for so long. At least one man was listening, and Thomas' words confirmed what he had long believed. George J. Snelus had carried out experiments with burnt lime linings in the early 1870s – his patent is dated 1872 – and though he had not pursued the developments further, because of other commitments, he now felt it prudent to do so.

The exposé of all this experimental work took place at the Paris meeting of the Iron and Steel Institute in the autumn of 1878. 'On the elimination of phosphorus' by Thomas and Gilchrist, subsequently published in the *Journal of the Iron and Steel Institute* (1879) was destined to become a metallurgical 'classic'. The cousins were followed by Snelus with his paper, 'On the removal of phosphorus and sulphur during the Bessemer and Siemens – Martin processes of steel manufacture,' and there were then no less than a further three papers all addressing the phosphorus problem.

Edgar Windsor Richards, Manager of the Bolckow Vaughan Works at Middlesbrough was present at the Paris meeting, and was introduced to Thomas. As a result, he visited Blaenavon Works in the company of Bolckow's consulting analytical chemist, John E. Stead. After seeing some trial blows Richards invited the cousins to continue their work at Bolckow's.

Henry Bolckow, the leading figure in Bolckow, Vaughan and Co Ltd. was an astute iron master who had long realised that the days of the wrought iron industry were numbered. Unlike Cleveland's other iron masters, who seemed powerless to resist the encroachment of Bessemer steel producers on their established iron rail market, Bolckow adopted the philosophy, 'if you can't beat them, join them.' At Bolckow Vaughan's Annual General Meeting in March 1871, he had told the company's shareholders that 'the establishment of a Bessemer Steel Works was almost a necessity.'

An order for a Bessemer converter had already been placed by this time, and matters took a slight twist when the company also decided to purchase the Bessemer works of the bankrupt Lancashire Steel Company at Gorton, near Manchester. Bolckow regarded the Gorton development as offering two principal advantages, even though, as he told shareholders, 'the works are not exactly in the position we would have wished.' The Gorton acid Bessemer plant could be operated profitably while Bolckow's Eston (Cleveland) Bessemer plant was being build, and it would also provide the company with knowledge of the steelmaking techniques required.

In 1872 a hematite ore mine in Northern Spain was acquired (jointly with John Brown and Company of Sheffield) to allow suitable acid Bessemer pig iron to be produced in Bolckow's Cleveland blast furnaces. Events at Gorton and Eston did not proceed as smoothly as desired. Gorton works had closed down by 1874 and Bolckow's shareholders, at that year's AGM, were informed of the intention 'to move the machinery to Eston.' Unspecified delays meant that commercial production of steel at Eston did not commence until the autumn of 1877. Consequently, when Thomas and Gilchrist transferred their work to the North East, acid Bessemer steelmaking on Teesside had been established barely a year.*

Two small Bessemer converters, capacity 1.5–1.75 tons, were erected by Bolckow Vaughan for the basic trials. Meantime, Richards had been empowered to negotiate terms for future development work whereby Bolckow Vaughan paid £1400 towards the expenses already incurred by Thomas and Gilchrist. In the event of successful conclusions to the work, Bolckow Vaughan was to have certain production rights and a lien over royalties from other steelmakers adopting the process.

Initial work at Bolckow's concentrated on producing a durable converter lining, which was finally achieved by binding burnt dolomite with tar. Even then, the problems were not over, as it had been found that phosphorus removal during blowing was erratic. Only after Stead pestered Richards to try an extended blow during conversion was efficient phosphorus removal achieved. On the 2nd April 1879 Thomas filed a British patent which specified this after-blow technique and protected the rights for it. Two days later Windsor Richards invited local ironmasters and capitalists to view the improved process: they witnessed the successful conversion of 1.5 tons of iron to steel in a little over 30 minutes.

The effect of the successful demonstration on the metallurgical world was related by Windsor Richards in his Presidential Address to the Cleveland Institution of Engineers on 15th November 1880. He told his audience, 'The news of this success spread rapidly far and wide, and Middlesbrough was soon besieged by the combined forces of Belgium, France, Prussia, Austria and America. The next meeting of the Iron and Steel Institute in London, under the presidency of Mr. Edward Williams, was perhaps the most interesting and brilliant ever held by that Institute. Directly the meeting was over, Middlesbrough was again besieged by a large array of Continental metallurgists and a few hundredweights of samples of basic brick, molten metal used, and steel produced, were taken away for searching analysis at home. Our Continental friends were of an inquisitive turn of mind, and like many other practical men who saw the process in operation, only believed in what they saw with their own eyes and felt with their own hands, and were not quite sure even then.'!

Later in his address Richards outlined the techniques employed at Cleveland Works in basic Bessemer steelmaking.

---

*There had been a notable attempt to introduce Bessemer steelmaking into the North East. Charles Attwood, a leading figure in the Tudhoe Iron works at Spennymoor, in County Durham, realised the potential of the Bessemer process after trial rolling some steel ingots at Tudhoe into ships' plate on Bessemer's behalf. Tudhoe at the time was a puddling and rolling plant sourced with pig iron smelted from Weardale ores at Tow Law blast furnaces, and Attwood arranged for 20 tons of Tow Law pig iron to be sent to Bessemer's Sheffield works for conversion to steel. Attwood and his manager witnessed the trial conversion and as a result he had Bessemer converters installed at Tudhoe about 1861. The plant consisted of four small converters (2.5 tons nominal capacity) mounted at the end of the arms of a cross which acted as a carousel or roundabout. This arrangement allowed each converter in turn to be positioned under the charging spout from a cupola and, after blowing, repositioned for tapping into the ladle. Some Bessemer steel was made at Tudhoe, but the venture was not a success. Instead, the Tudhoe rolling mills were largely sourced with steel ingots from an early Siemens open hearth plant installed at Wolsingham, near the Tow Law blast furnaces. The Tudhoe Bessemer plant was derelict by 1890, and the site cleared about 1920.

'Our plan of operation is exceedingly simple. The converter, as is usual, is first heated up with coke, so as to prevent the chilling of the metal; then a measured quantity of well-burnt lime, about 16 per cent. of the weight of molten metal, mixed with a small quantity of coal or coke, is charged into the converter and blown till the lime is well heated. The molten metal is then poured on the lime additions, the blast of 25 lbs. pressure is turned on, and the carbon lines disappear in about ten minutes; then, after about 2 1/2 minutes over-blow, the converter is turned down and a small sample ingot made, which is quickly beaten into a thin sheet under a small steam hammer, cooled in water, broken in two pieces, and the fracture shows to the experienced eye whether the metal is sufficiently ductile. If it is not so, then the blowing is prolonged, after which the spiegel is added, and is now being poured into the ladle, not into the converter. For the basic process the initial bath should be low in silicon, because silicon fluxes and destroys the lining, and causes waste of metal; it should be low in sulphur, so that the metal may not be red-short. Nearly one-half the sulphur is eliminated by the basic process. In order to work economically, the metal should be taken direct from the blast-furnace, so as to avoid, first, the cost of re-melting in a cupola, and, second, to avoid further contact of the metal with the sulphur and impurities of the coke. It is not an easy matter to accomplish in a blast-furnace the manufacture of a metal low in silicon and, at the same time, low in sulphur. It would, no doubt, very much help to keep sulphur low if manganese were added in the blast-furnace; but manganese is a costly metal. At present we have succeeded in making a mottled Cleveland iron with 1% silicon and 0.16% sulphur, and white iron with 0.5% silicon and 0.25% sulphur, which, taken direct from the blast-furnace, have both made excellent steel.'

Another method described by Richards – the 'transfer' system – enabled grey iron to be taken from the blast-furnaces to the converter without considering the percentage of sulphur.

'But we have another method of operating which relieves us from the necessity of making a particular quality of Cleveland pig iron. We call this second mode of working the 'transfer' system. because we transfer the metal from the acid to the basic converter. The transfer system enables us to take any grey iron direct from the blast furnace without any consideration as to the percentage of sulphur, which is always low in grey iron. This grey metal is poured into a converter with a siliceous lining and desiliconised, when, after say 12 to 15 minutes' blowing in the ordinary manner, it is poured out of the converter into the ladle, and poured again from that ladle into a converter lined with dolomite, taking care that the highly siliceous slag is prevented from entering the basic lined converter. Then in the second converter it is only necessary to add sufficient lime for the absorption of the phosphorus of the metal and blowing then need not occupy more time than is necessary for the elimination of the phosphorus, say about three minutes. This mode of operation will, no doubt, give the basic lining and bottom a much longer life, but we have not yet been long enough at work to obtain the necessary experience to determine which is the better system of working, but both are good and effective and have given excellent results.'

Richard's 'transfer' system was nothing short of 'duplex' working, which was first described to the Iron and Steel Institute at the 1878 Paris meeting by Harmet, and was subsequently adopted much more widely as a steelmaking technique.

The success of the pioneering basic steelmaking at Bolckow Vaughan's had far-reaching implications. Not only did it mean that local Cleveland ores could finally be used to make steel: it paved the way for a whole new European steel industry based on the abundant phosphoric minette ores of Lorraine and Luxembourg.

BASIC BESSEMER WORKS

ESTON/CLEVELAND WORKS

Following on from its initial experience at Gorton, Bolckow Vaughan installed four 8 ton acid converters at Eston, and began operating them in 1877. The output was essentially all rolled to rails. After the successful trial work of Thomas and Gilchrist, one or two of these vessels were operated in 1879/80 with a dolomite lining to prove the commercial viability of the basic process.

In the early 1880s Bolckow Vaughan erected a new steelmaking plant at Eston, which included six basic lined converters. With an individual capacity of 15 tons the vessels were considerably bigger than the company's earlier converters, and the shape of the mouth was altered to take account of the slag behaviour during blowing. (The symmetrical converter mouth which was found to be the most effective was actually developed by Holland and Cooper working at Brown, Bayley and Dixon.) The converters were installed piecemeal in pairs in 1880, 1881 and 1882, and mounted on a raised platform so that ingot teeming could take place at ground level rather than in pits. The ingots were rolled variously to rails, plates, angles, sleepers, pit rails, fish plates and merchant bar.

Two related techniques were adopted in an attempt to ensure that acceptable steel could be produced more consistently. Hot metal mixers were installed in 1893 to reduce the variation in composition of the iron fed to the converters, and the Massenez Process of charging iron oxide (ore) to the converter adopted. The latter was claimed to reduce slopping from the converter during blowing, and to work effectively with a range of pig irons. Despite these steps, however, the company still experienced difficulty in producing acceptable steel consistently because of the wide variations in blast furnace metal composition, and all the Eston Bessemer converters were closed down in 1911, being superseded by open hearth steelmaking.

NORTH-EASTERN STEEL

The North-Eastern Steel Co Ltd was registered on 9th July 1881 'to carry on business as steel converters and steel and iron manufacturers.' The first directors were R. C. Denton, A. J. Dorman and T. Wrightson, though Sidney Thomas and Percy Gilchrist were amongst the founders.

Land was obtained in the 'Ironmasters District' north west of Middlesbrough railway station on the south bank of the River Tees and plans for the works were drawn up by the newly appointed Manager, Arthur Cooper (formerly of Brown, Bayley and Dixon).

Orders for the plant had been placed by the end of 1881, and the first cast of steel produced on 31st May 1883. As at Eston, the practise of mounting the converters on a raised platform was adopted, and the four 10 ton converters were operated on a cycle of three working and one off for relining/repair. Arrangements were concluded with neighbouring blast furnace operators for a supply of suitable hot metal in rail-borne ladle bogies for the converters. However, trials showed that this arrangement was not practical as the works was originally laid out, and only when a mixer furnace was installed in 1893 could local hot metal be utilised as feedstock. In the meantime, cupolas were employed to provide liquid iron as well as liquid spiegel for the converters. Cooper got the idea for a metal mixer from David Evans, who had been General Manager of Barrow Haematite Steel Co Ltd, and came to Teesside in 1892 as newly appointed Works Manager at Bolckow Vaughan. (Evans' arrival also explains the adoption of mixers at Eston at this time.)

As laid out, the works was capable of producing 2 000 tons per week of steel products, with the principal output consisting of rails, blooms, billets and bars. End uses for the 'semis' included sleepers, tinplates, stampings, wire nail strips, tubes etc.

In 1899 the company introduced the then novel system of recarburising the blown charge with anthracite additions rather than just relying on the carbon present in the spiegel to bring the composition back into specification.

North-Eastern Steel became a subsidiary of Dorman Long & Co Ltd in 1903 and Bessemer steelmaking was retained until 1919.

ERIMUS/DARLINGTON/GUISBOROUGH

Three other North Eastern concerns also expressed an intention to use the basic Bessemer process but changing circumstances prevented this. Possibly the most interesting of these was the Erimus Steel Works, located near Newport (west of Middlesbrough) where a Bessemer works was built in 1880/81 on the site of a former wrought iron works. Equipped with two 6 ton converters, it was capable of producing 600–700 tons of rails per week. Though the stated intention was to adopt the basic process, only hematite pig iron was used. Some provision, however, had been made to treat phosphorus-rich pig iron, with the plant laid out in such a manner that a desiliconised (i.e. part-blown) charge from an acid converter could be transferred to a second converter (basic-lined) for duplex working. At Darlington, a Bessemer plant with two 7 ton converters was recorded as under construction in 1880 by the Darlington Iron Company, a business started in the 1860s by William Barningham. The stated intention of basic working was, as at Erimus, never brought to fruition and the acid process was employed, using hematite pig iron remelted in cupolas. *Iron,* issue dated 2nd November 1883, recorded the 1882 output of steel rails at 40 000 tons. The undertaking traded as the Darlington Steel & Iron Co Ltd from 1882 until its liquidation in 1897/98. Part of the site was then taken over by the Darlington Forge Co Ltd, which itself became a subsidiary of the English Steel Corporation in 1936. Finally, there was a plan to install Bessemer converters (presumably basic lined) at the Cleveland Iron & Steel Works, Guisborough, a foundry business which by the 1880s was being run by partners John Sutherst and William Southorn. In the event, the project was still-born though much later, in 1917, the successor company (Blackett, Hutton & Co) installed a 3 ton Stock Converter fired by fuel oil, in which hematite pig iron was melted 'in situ' before being blown down to steel. This continued to operate until the early 1950s.

BROWN, BAYLEY & DIXON

Of all the Sheffield district acid Bessemer steel railmakers, only Brown, Bayley & Dixon directly addressed the growing problem of raw materials costs, which were increasingly putting the inland rail producers at a disadvantage to coastal plants.

There seems little doubt that the company had realised by 1879 that a successful commercialisation of the basic process, then under development, would offer such a solution by allowing cheaper pig iron to be used by existing Bessemer operators.

A year earlier, in 1878, the company was actively engaged in helping to solve the problem of finding a serviceable basic lining for the Bessemer converter. Consulting chemist Edward Riley, while working for the firm, took out a patent in November 1878 for the use of oil as a binder for dolomite furnace linings. The first tentative steps towards commercial production of basic Bessemer steel had taken place by 1879. According to C. B. Holland and A. Cooper, a technique they had witnessed at the Hoerde Company in Westphalia (Germany) to prevent phosphorus reversion was commenced at Sheffield on 20th November 1879. The pair mentioned this in a paper to the Iron and Steel Institute in 1880, and in the same volume R. Pink of the Hoerde Company provided the technicalities. Following the afterblow, which took the phosphorus into the slag, additions of spiegel were deferred until most of the slag had been physically skimmed off the melt; simple yet effective. It is heartening to note that such free interchange of information was already being practised enthusiastically by Institute members: Pink's paper followed the visit of Hoerde representatives to the public demonstrations of basic Bessemer steelmaking given at Middlesbrough (Eston) in 1879, when a goodly number of continental metallurgists had been present. Holland and Cooper were also responsible for developing a modified form of converter mouth which was able to cope more effectively with the increased slag bulk of the basic process.

Matters looked particularly promising by the spring of 1880, when Brown, Bayley & Dixon made an average of 422 tons of basic steel a week over a two month period. Unfortunately, the technical success was not matched by the company's financial situation, and the business failed early in 1881. It is said that this was brought about because of a heavy loss made on a rail order to Russia supplied at a price which didn't cover the cost of the pig iron. How much of this could be laid at the door of Dixon, one of the partners, is open to speculation. Dixon, before joining the venture, had been the St Petersburg representative of John Brown & Co Ltd of Sheffield, and considered something of an expert on trade with Russia. Possibly the company traded over-zealously with Russia in an attempt to fill a dwindling order book, and accepted contracts at unrealistic prices. Perhaps it is significant that when the new company was constituted, Dixon's name was absent and the business traded simply as Brown Bayley's Steel Works Ltd. The new company gave up unprofitable rail rolling and concentrated on the production of railway materials – tyres, axles, springs – and steel 'semis' (blooms, billets and bars). Arthur Cooper, one of the protagonists of the basic process, left in 1881 to take up the position of manager at the newly formed North-Eastern Steel Co Ltd, and further work on the basic process at Sheffield was suspended. Acid Bessemer steel continued to be produced (alongside open hearth steel) and the last blow was made in January 1918.

THE LILLESHALL COMPANY

The Lilleshall Company grew from the desire of the Second Earl Gower (Granville Leveson-Gower) to exploit the rich mineral deposits on his lands in Shropshire. By the late 19th Century it had grown into a large complex of coal and ironstone mines; blast furnaces plants; puddling furnaces, forges and rolling mills for wrought iron production; and a heavy engineering business.

Lilleshall installed a basic Bessemer steelworks in 1882 at its Priors Lee Iron Works site (now a suburb of Telford). The plant is said to have been built under the personal direction of P. C. Gilchrist. It was equipped with three 7 ton converters, supplied with hot metal from the adjacent blast furnace plant, and produced some 700 tons of steel ingots per week, working on a two 12 hour shift pattern. The ingots, nominally 10–12 cwt weight, were rolled to blooms and billets for subsequent re-rolling to bars and sections.

By the time Lilleshall introduced steelmaking, it was already an important supplier of iron- and steelworks plant and ancillaries. Under the circumstances it is highly likely to have made the converters and other equipment for its own steelworks. Certainly it was supplying such equipment to others, for example the Leeds Steel Works (q.v.), by the 1890s.

The Lilleshall Bessemer plant, together with a later (c1900) Siemens basic open hearth furnace, closed down in the early 1920s because of the deteriorating economic climate, and thereafter the company bought in steel billets for re-rolling.

GLENGARNOCK WORKS

Glengarnock Works was started in 1843 by partners Merry and Cunninghame with the erection of eight blast furnaces to produce pig iron from local ironstone deposits. Situated northeast of Ardrossan, and only some ten miles from the Ayrshire coast, the plant was rather remote from the Clyde Basin, where its principal customers were situated.

In 1884 the works expanded into steel production by the installation of four 8–10 ton (nominal) basic Bessemer converters, along with ancillaries which included a 10 ton steam hammer and cogging mill. The converters were supplied with hot metal from the adjacent blast furnace plant. The initial intention was to concentrate on the manufacture of steel plates, angles and tinplates blooms but various factors, particularly the suspicion within the shipbuilding industries for Bessemer steel, mitigated against this. As a result, the company began the production of joists or 'H' beams for structural and bridge building purposes.

Steelmaking capacity was increased in 1892, but now utilising three 25 ton capacity open hearth furnaces, and obviously with an eye to providing 'acceptable' steel to the Clyde shipyards.

After standing idle due to a trade depression, Glengarnock Works was restarted when it was rented to David Colville & Sons Ltd in 1915 as a direct result of intervention by the Ministry of Munitions, keen to maximise the UK's iron and steel production for the war effort. Colville's purchased the plant outright in 1916 and continued to work the Bessemer furnaces until 1920, after which Glengarnock only made steel by the open hearth process.

BILSTON WORKS

South Staffordshire was an area where iron working, especially the production of wrought iron and its products, was long established and very widespread. By 1880, for example, the area's wrought iron industry boasted 1625 puddling furnaces and 311 rolling mills within 125 works. These were sourced largely from open top blast furnaces smelting local ores, although in 1880 only 45 of the area's 137 blast furnaces were in blast. Little wonder the district had become known as the Black Country.

Bilston, south east of Wolverhampton, grew up as an industrial complex within which were several collieries, blast furnaces and iron works (i.e. puddling and rolling plants). Alfred Hickman acquired Bilston Furnaces (blast furnaces) about 1866, renaming them the Springvale Furnaces. In 1882 he promoted the formation of the Staffordshire Steel and Ingot Iron Co Ltd to manufacture steel by the basic process.* Second hand Bessemer equipment – three 12 ton (nominal) converters and ancillaries – was obtained from the Mersey Steel and Iron Co of Liverpool and installed in 1886. P. C. Gilchrist was serving as a director of the Bilston business at about this time, and it later traded under various titles which included the name of the founder. In 1920 a controlling interest in Sir Alfred Hickman Ltd (as it was then known) was obtained by Stewarts & Lloyds Ltd. Basic Bessemer steelmaking continued until 1925, after which all steelmaking at Bilston was carried out in openhearth furnaces.

It is relevant at this point to note that a significant secondary industry grew up following the coming of basic Bessemer steelmaking. Initially the phosphorus-rich basic slag from the process seemed to prove something of an embarrassment on Teesside, and whilst some was shipped to Germany the remainder was simply dumped in the North Sea. Bilston, however, was more alive to the potential of this voluminous by-product, and as early as 1886 announced its intention to install a slag grinding mill. Crushed basic slag was soon being used as an artificial manure or fertiliser in agriculture, where its slow release of phosphorus into the soil helped to improve crop yields.

LEEDS STEEL WORKS

By the middle of the last century the Leeds-Bradford area was firmly established as an area of excellence for wrought iron production and products, especially railway materials (components for locomotives and rolling stock). 'Best Yorkshire' was a marque for such goods and fully the equal of 'Made in Sheffield' on edge tools and cutlery.

A 'club' had gradually evolved consisting of a tightly-knit group of producers who zealously guarded their reputation by 'policing' their raw materials rigorously and sticking to well-proven production routes. Crucial to their success was the use of local coal-measure

*Hickman was not the first ironmaster in the Black Country to install Bessemer converters. Lloyd, Foster and Company at Wednesbury had two 3 ton acid converters by 1867. This undertaking, later trading as the Patent Shaft and Axletree Company, provided facilities in 1881 which allowed Hickman to assess the potential of the Bessemer process (in conjunction with a basic lining) for converting high phosphorus pig iron into steel. Hickman's conclusions, summarised in the *Journal of the Iron and Steel Institute* (1882, Part I, pages 195–196) convinced him of the viability of the process and led directly to the formation of the Staffordshire Steel and Ingot Iron Co Ltd.

ores, low sulphur coals and the sole employment of pig iron smelted in open top blast furnaces equipped with cold blast. Bowling, Low Moor and Farnley were some better known names associated with the trade.

There were iron smelters in the area supplying merchant grades of pig iron, but they were in the minority. One of these merchant pig producers, the Aireside Hematite Iron Company, situated in the Hunslet district of Leeds, would provide the rather unlikely setting for a basic Bessemer steelplant (Fig. 1).

Founded by partners Ledger and Cooper, the works originally consisted of three blast furnaces, but by 1886 (per *Iron* of 9th April 1886) the owners were contemplating the manufacture of steel. This desire to move up market came closer to fruition on 26th June that same year with the registration of the Aireside Steel & Iron Company Ltd; the company prospectus gave its objectives as acquisition of the property of Aireside Hematite Iron Co and the addition of a steelworks. One of the principal instigators was railway contractor Walter Scott of Newcastle.

For some reason (possibly undersubscription) this new company was superseded by another, the Leeds Steel Works Co Ltd, a private company registered on 29th October 1888. Walter Scott, John Scott, Robert Hammond and Lord William Beauchamp Nevill were the first directors of the new company. Somewhat later, on 4th December 1900, the assets of the Leeds Steel Works Ltd were amalgamated with Walter Scott's other industrial undertakings (principally colliery properties in Durham) and the new group traded as Walter Scott Limited.

The Bessemer plant and steelworks were installed about 1890: Griffiths *Iron and Steel Manufacturers* of 1891 lists the plant with two converters at that stage. During a visit to Leeds in July 1903 by the Institution of Mechanical Engineers, the steelmaking plant consisted of four basic Bessemer converters (two at 7 tons capacity; two at 10 tons) served by two semi-circular casting pits and hydraulic cranes for the casting ladle and ingot moulds. The later converters (at least) were supplied by the Lilleshall Company of Shropshire, which described them as 15 ton (nominal). Most of the iron feedstock was brought directly to the Bessemer department as 'hot metal' in 12 ton capacity ladles carried on four-wheel rail bogies, and stored until required in a 120 ton unfired mixer. Three cupolas were also available to provide an alternative source of liquid pig iron.

Interestingly, the blast furnace plant was fed with a mixture of Lincolnshire and Northamptonshire ores, together with puddling furnace 'tap cinder' (slag rich in iron and phosphorus). It should be remembered that this 'adulteration' of the blast furnace burden by puddling furnace slag was quite a widespread practice. In particular its adoption in Cumberland resulted in the production of high phosphorus pig iron ostensibly from hematite ores, and this had caused Bessemer a few problems in the early days of his 'new' steelmaking process when he was taking all pains to ensure the 'pedigree' of the pig iron feedstock.

The Leeds Steel Works produced ingots weighing from 1 ton to 2 tons each and these were rolled initially through a 28 inch cogging mill and then a 32 inch roughing and finishing mill. The principal output consisted of steel girders, joists, channels, flats and tram rails for street tramways. At the time of the visit by the Institution of Mechanical Engineers the works was offering up to 63 sections of joists and channels and 44 tram-rail sections. The latter were of a girder section and Leeds Steel Works claimed to be the largest makers of such

Fig. 1   Bessemer converters in Leeds Steel Works with broadbent overhead crane.
(Courtesy T.J. Lodge)

items in Great Britain. Demand for longer rails had resulted in the rolling of 60 foot lengths, the first of these being supplied to Glasgow Corporation in 1899.

In some ways the specialism in street tramway rails became an 'Achilles Heel' for the company particularly when municipal authorities began to favour the introduction of motor buses rather than extend tramway routes into new outlying suburbs. Attempts were made to diversify production following the Great War, and steel 'semis' – blooms, billets, slabs – feature on the company's list of products for the mid 1920s. Manufacture of railway rails also commenced, with the company joining the Rail Manufacturers' Association in August 1921. Such moves, however, were not enough. Bessemer steelmaking ceased about 1925 and latterly the works specialised in supplying various grades of iron for foundry and forge work and basic iron for other steel works. The works suffered complete closure about 1930 as yet another victim of the 1920s Depression.

CORBY WORKS

The original company owing the Corby steelworks, Stewarts & Lloyds Ltd, was formed in 1903 by an amalgamation of the tube making businesses of Lloyd & Lloyd Ltd and A. & J.

Stewart and Menzies Ltd. the largest tube makers in England and Scotland respectively. The new business consisted of a large number of diverse plants scattered through the West Midlands and the Clyde Basin, and producing a bewildering range of products. Initially the business lacked focus, and so often did the two 'halves' seem to be at variance that one 'wag' was provoked to describe the company as 'Stewarts against Lloyds'!

It was soon apparent that the company had to acquire its own iron- and steelmaking capacity, and to rationalise production if it was to make the business more efficient. After some initial work in this direction the company was finally able to achieve all these objectives on a grand scale with the building of Corby Works in Northamptonshire.

The Corby scheme was one of several bold initiatives considered in the 1930s with a view to carrying through a whole range of sweeping rationalisations within the stagnant UK iron and steel industry. Government incentives were provided as an encouragement, usually with the Bank of England acting as broker and adviser to the various schemes proffered by the steel barons.

Stewarts & Lloyds had already had a survey carried out by American consultants H. A. Brassert & Co Ltd as to the best option for rationalising its business so that it could compete effectively in world markets. Brassert's recommended that the existing tube works, 'small and scattered', be integrated on one site. It concluded that Corby, surrounded by deposits of phosphoric Northamptonshire iron ore, had marked advantage for the manufacture of both basic Bessemer and open hearth steels to support such a venture. Significantly, perhaps, Stewarts & Lloyds was already using continental Thomas (i.e. basic Bessemer) steel for lapwelded tubes produced at its Phoenix Works in Coatbridge.

As conceived by Brassert's, Corby Works would be able to make basic Bessemer steel and convert it to tubes in a complex 'equal to the best practice on the Continent'. Brassert's recommendations for the integrated iron and steel complex included three 25 ton Bessemer converters. Output would be such that after conversion there would be surplus sheet, bars and billets available for outside sale. It was anticipated these would be sold to inland UK customers (principally in the Birmingham area) who at that time were purchasing continental Thomas steel.

Once the scheme was finally sanctioned, the site for the basic Bessemer plant was made ready by September 1933. Because there was no 'modern' Bessemer steelplant then existing in the UK, it was necessary to go to the continent to obtain the necessary equipment. The blowing vessels and ancillaries came from the German Gutehoffnungshütte company, supplied through agents Pearson and Knowles of Warrington. The plant included a 1 000 ton metal mixer to hold the hot metal from three new blast furnaces erected on site (one of which was the largest in the UK).

Corby made its first steel on 27th December 1934. Reporting on the development, *The Times* for 8th January 1935 said,

'Now and then a great commercial or industrial undertaking takes hold of the mind of a people and becomes a symbol of enterprise, high courage and progress. Corby typifies the spirit of industrial resurgence. Today from what was until recently a quiet village in the heart of the Midlands comes the news that the manufacture of Basic Bessemer steel has been restarted in this country'.

A useful technical description of basic Bessemer steelmaking at Corby is given in the Iron & Steel Institute's Special Report No 42 *Report on the Bessemer Process* (1949). By this time Corby was operating with three 1000 ton metal mixers and five 25 ton Bessemer converters. The plant only blew one vessel at a time, and visual observation and spoon samples from the converter were sufficient to monitor the blow – no spectrometer or photo cell was used. At the conclusion of the blow, the slag was skimmed off and ferro manganese added to the vessel. This was then held for three minutes before the steel was poured into the ladle, when aluminium additions were made. Ingots of 4 tons weight were produced by top pouring (ie direct teeming) and further aluminium was added only under certain circumstances (eg to control rimming action). To produce lower nitrogen steels suitable for tube making, the converters had been widened to give a shallow bath of metal, and additions of ore or mill scale were made, rather than scrap.

The closing comments of those responsible for the Special Report are most illuminating. The sub committee admitted that some of its members who previously 'had felt somewhat sceptical about the usefulness of the Bessemer process, came to the unanimous conclusion that the thorough and regular control of present-day Bessemer practice results in a regular output of high-quality steel which contrasts greatly with the product of the irregularly operated process of bygone years'. A clear stamp of approval undoubtedly, but only a stay of execution.

In 1962 successful steelmaking trials were carried out at Corby with a 100 ton prototype Basic Oxygen Steelmaking (BOS) vessel, and as a result a contract was placed for extensions to the steelplant which would create a three BOS vessel facility to replace the basic Bessemer converters. The new plant, costing £6.9 millions, was commissioned in July 1965 and the Bessemer converters fully superseded in January 1966.

SOURCES/BIBLIOGRAPHY

In addition to the specific references quoted in the text of this chapter, the following titles have been consulted.

K. Barraclough: *Steelmaking 1850–1950,* Institute of Metals, 1990.
W.K.V. Gale and C.R. Nicholls: *The Lilleshall Company Ltd.: A History 1764–1964,* Moorland Publishing Co. Ltd., 1979.
C.A. Hempstead (ed.): *Cleveland Iron and Steel,* British Steel Corporation, 1979.
P.L. Payne: *Colvilles and the Scottish Steel Industry,* Clarendon Press, Oxford, 1979.
Sir Frederick Scopes: *The Development of Corby Works,* Stewarts & Lloyds, 1968.

I have also used selected information from my own unpublished manuscripts on the histories of Brown, Bayley & Dixon; Alfred Hickman (Bilston); and the Leeds Steel Works.

Impact of Bessemer Steelmaking Technology on Industrial Growth in the United States and the Subsequent Evolutions in Steelmaking Processes

E.T. Turkdogan

During the past Century many papers and books were published as a tribute to Sir Henry Bessemer, the prolific inventor of a wide range of manufacturing processes. The Bessemer process of steel production was of course his most important invention that provided the much needed material at the onset of the rapidly expanding industrial developments in engineering, railways and shipbuilding.

INTRODUCTION

In 1850 the U.S. Congress began granting federal land to all states for the development of transcontinental railways, which was the beginning of a major industrial boom. In the mid nineteenth Century, steelmaking in the United States was in its infancy. It was 'wrought iron,' produced from pig iron in a puddling furnace that was used in making rails at the rolling mills. One major problem in the developing railroad industry was the wear of wrought iron rails which needed frequent replacement. Yet in the early 1860s, steel rails were used successfully in Europe using the Bessemer steel that was produced mainly in Great Britain.

As outlined briefly in this paper, the Bessemer steelmaking technology had a major impact on the industrial growth in the United States of America and also on the subsequent evolutions in steelmaking processes.

START OF BESSEMER STEELMAKING IN THE USA

When Bessemer was developing his pneumatic steelmaking process in the mid 1850s, there were also other inventors, e.g. Mushet, Martien and in particular Kelly, experimenting to refine pig iron to malleable iron by blowing steam or air into the puddling furnace. Kelly, an ironmaker in the Southern States of America, was using an air blast to refine pig iron some six or seven years before Bessemer became interested in steelmaking. Although iron rails made by Kelly's 'air boiling' process had been used for some time, the process was ultimately abandoned. What was not achieved in Kelly's initial air boiling process was melting of the purified iron. It was only after Kelly patented his process in June 23, 1857, subsequent to Bessemer's UK key patent No. 356 dated 12th February 1856, that he achieved a liquid product in his air boiling process. His intention was primarily to improve the puddling process in converting pig iron to malleable iron. In contrast, Bessemer's intention was to make a better product (steel) than that obtained by the puddling process. It was Bessemer's technical ingenuity and his business associations in various industrial developments both in Great Britain and Europe that made his pneumatic steelmaking process prevail on Continents both sides of the Atlantic.

Although in violation of Bessemer's machinery patents, the first experimental Bessemer converter of 2.5 ton capacity was built in the United States at Wyandotte near Detroit, Michigan for the Eureka Iron Company owned by Eber Ward. This plant was in operation by September 1864 but was subsequently abandoned in 1869. An account of this experimental Bessemer plant was later documented in a paper by Durfee.[1]

Alexander L. Holley, a talented consultant mechanical engineer and author of many publications on engineering subjects, played a key role in implementing and advancing the Bessemer steelmaking technology in the United States. While holding a position at the Camden & Amboy Railroads, in 1862 he was assigned to gather technical information on shipbuilding, armour plates and armaments in Europe. During this time, he visited the Bessemer Works in Sheffield. In his book on this European trip, Holley[2] summarised his observations by the following statement. 'The Bessemer process of making steel promises to ameliorate the whole subject of ordinance and engineering construction in general, both as to quality and cost.'

On his return from Europe, Holley was provided with financial support by the ironmaster John F. Winslow and banker John A. Griswold to undertake a steelmaking venture. Upon securing an American license for Bessemer's patents in 1864, Winslow, Griswold and Holley in partnership commenced planning a Bessemer plant to be installed at Troy, N.Y. The 1865 plan of the Troy Bessemer plant, with an experimental 2 ton converter, is shown in Fig. 1, reproduced from a paper by Hunt.[3] The first experimental converter blow was made on 16th February 1865. Although the first Bessemer steel was made at an earlier date (6th September 1864) at the Wyandotte Works, the Troy establishment engineered by Holley, was the first to bring the Bessemer process near to a commercial success. Two years later, Holley extended the Bessemer plant and increased the converter capacity to 5 ton.

ESTABLISHMENT OF BESSEMER ASSOCIATION

Growth of steel production in the United States was initially structured and monitored almost entirely by the railroad companies. In the early 1860s the rail mills organised patent pools incorporating the Bessemer technology, which ultimately led to the formation of the Bessemer Association. However, there were patent litigations between two competing groups. Eber Ward, owner of the Eureka Iron Company, controlled Kelly's patent and the American rights of Robert Mushet's patents for treating steel with spiegeleisen to improve its mechanical properties. The Troy group of Winslow, Griswold and Holley held the American rights of the Bessemer patents. Litigation of a formidable nature was ultimately settled out of court by 1869, which resulted in pooling the Kelly and Bessemer patents. The Troy group received 70 percent of the proceeds from licensing and the Ward group 30 percent. These events were

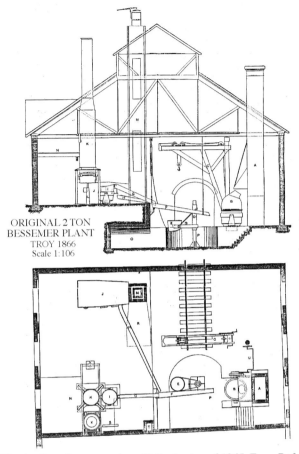

ORIGINAL 2 TON
BESSEMER PLANT
TROY 1866
Scale 1:106

Fig. 1   Troy Bessemer plant, Holley's plan of 1865. From Ref. 3.

later reported by Holley[4] in a paper entitled 'The invention of the Bessemer Process.' Similar accounts of these events have been documented in several subsequent publications on the history of the steel industry in the United States of America. Cited references 5, 6 and 7 are for more recent publications on this subject.

For the administration of the patent pool the following organisations were established: Pneumatic Steel Association, the Bessemer Steel Association and the Bessemer Steel Company. From 1866 to 1877 the Bessemer Association licensed eleven plants. As noted in Table 1, of the thirteen Bessemer plants built between 1864 and 1876, eleven of them were designed by Holley.

To minimise competition, the Bessemer Association severely restricted the number of licenses after 1877 and also restricted the circulation of reports on the art of Bessemer technology only to association members. Furthermore, from 1877 to 1915 the price of steel rails was mainly determined by the Bessemer Association. In the 1880s steel rails were sold at about $47 per tonne, decreasing to $31 per tonne by the 1910s.

Table 1   Initial Bessemer plants built in the United States during the period 1864–1876. From Ref. 3.

First Blow	Company	Location	Type of mill
September 1864	Eureka Iron Company*	Wyandotte, Mich.	Rolling Mill
February 1865	Winslow & Griswold	Troy, N.Y.	Rolling Mill
June 1867	Pennsylvania Steel	Harrisburg, Pa.	Railroad
May 1868	Freedom Iron & Steel*	Lewistown, Pa.	Rolling Mill
October 1868	Cleveland Rolling Mill	Newburgh, Ohio	Rolling Mill
July 1871	Cambria Iron	Johnstown, Pa.	Rolling Mill
July 1871	Union Iron	S. Chicago, Ill.	Rolling Mill
April 1872	N. Chicago Rolling Mill	Chicago, Ill.	Rolling Mill
March 1873	Joliet Iron & Steel	Joliet, Ill.	Rolling Mill
October 1973	Bethlehem Iron	Bethlehem, Pa.	Iron mill
August 1875	Edgar Thomson Steel	Pittsburgh, Pa.	New mill
October 1875	Lackawanna Iron & Steel	Scranton, Pa.	Iron mill
1876	Vulcan Iron	St. Louis, Mo.	Iron mill

*With the exception of these two mills, the other eleven plants were designed by Holley.

DEBATE ON THE DEFINITION OF STEEL

Over a period of twenty years from 1870 to 1890 there were heated debates on the definition of steel. The steel makers and users were advocates of the fusion classification as articulated by Alexander Holley in his paper entitled 'Bessemer Machinery':[8] 'Steel is an alloy of iron which is cast while in a fluid state into a malleable ingot.' This fusion classification would imply that if the metal had been completely molten it would become 'steel' regardless of its carbon content; if not it would remain 'iron.'

Metallurgists to the contrary insisted on the carbon classification as stated by Henry M. Howe:[9]

'The steel is a compound or alloy of iron whose modulus of resilience can be rendered, by proper mechanical treatment, as great as that of a compound of 99.7 per cent with 0.3 per cent carbon can be by tempering.'

He attacked the fusion classification on the grounds that 'all products of the Bessemer converter and open-hearth were labelled steel without regard to carbon content or mechanical properties.'

Holley strongly rebutted Howe's point of view in his paper entitled '*What is Steel*';[10] his argument being;

> 'The general usage of engineers, manufacturers and merchants is gradually, but surely, fixing the answer to this question... While the fusion classification was already in place, the carbon classification was arbitrarily devised and must bear the demerit... of upsetting existing order and development.'

Holley gained the support of industrialists and by 1880 the definitions of 'iron' and 'steel' were stabilised thus, 'steel' was liquid and pourable from the Bessemer converter, open-hearth and crucible furnaces. Wrought 'iron' was a spongy product of the puddling or boiling process, which was contaminated with slag, cinders and carbon flecks. Ultimately, however, Howe's carbon classification became the universally accepted definition of steel.[11]

EDGAR THOMSON WORKS OF ANDREW CARNEGIE

In 1865 Andrew Carnegie had resigned from the Pennsylvania Railroad (PRR) after gaining excellent managerial experience during his twelve years with the company. With the financial backing by Thomas A. Scott and J. Edgar Thomson, Carnegie soon became a successful entrepreneur in the iron and steel related ventures such as (i) Union Iron Mills for rail making, (ii) Keystone Bridge Company and (iii) Central Transportation Sleeping Car Company.

On behalf of PRR, in 1872 Carnegie went to England to investigate the Bessemer converter operation. He was much impressed with what he observed and was quickly transformed from a wrought iron man to a steel man. Recognising the need for domestic steel production in an ever expanding railroad construction, Carnegie formed a partnership with Edgar Thomson, Thomas Scott and others to build a new mill based on the Bessemer technology.

Carnegie had long recognised, from his previous experiences at PRR, the importance of employing the right people and organising them to ensure success in a challenging business venture. He stated in one of his writings;

> 'I am neither mechanic nor engineer nor am I scientific. The fact is that I don't amount to anything in any industrial department. I seem to have had a knack of utilising those that do know better than myself.'

Since Carnegie wanted a new plant to produce world class steel rails, he hired A.L. Holley (the most talented and experienced Bessemer engineer in the United States) to design and build his new steel mill. To ascertain a successful operation of the mill he also wanted to staff his plant with the best available engineers and managers. His wish soon materialised from a major labour dispute that took place in 1873 at the Cambria Iron Works in Johnstown, Pa. The disgruntled heads of departments at Cambria became available for hire by Carnegie for his plant under construction. He first hired William R. Jones (known as Captain Jones), who was an excellent plant manager and talented steelmaker; soon after, Carnegie hired most of the other experienced engineers and department managers from the Cambria Iron Works.

It took about two years to complete the construction of Carnegie's steel mill, which he named 'Edgar Thomson Works,' (E.T.) situated at the 106 acre site along the Monongahela River in the Braddock district of Pittsburgh, Pa. To speed up the plant construction, Carnegie and Holley contacted Junius S. Morgan, the international banker in London in the summer of 1874, to secure a $400 000 bond issue. The plant was operational by the summer of 1875 with the first blow in the Bessemer converter on 22nd August and the first rail production on 1st September.

It may be of some side line interest to note that an Irish crew in the plant controlled the Bessemer converter shop which was for decades referred to as the 'Kelly Converter' after William Kelly, the Pittsburgh inventor who was originally from Ireland.

Holley incorporated numerous engineering developments in his design of the Bessemer plant. For example, two 6 ton converters (6 ft i.d., 15 ft high) were installed with his patented detachable bottom. After eight blows, the used bottom block was unbolted and replaced with a new one while the converter was still hot (Fig. 2).[13]

In 1879 Holley constructed a heat exchanger at E.T. that used hot converter off gases to preheat an incoming air blast for the cupola furnaces for melting pig iron. An array of butterfly valves directed ambient air into the twin heat exchangers and returned the hot air blast to any of the three cupola furnaces. This arrangement made it possible to transfer heat continuously from the converters to the cupola furnaces, which was the initial step in converting the E.T. Bessemer plant from a batch to a continuous process.

William Jones, who had many years of experience in making iron and steel at various iron works, brought about a series of innovations at E.T. One of his inventions of major importance was the 'metal mixer,' intended for a fully continuous Bessemer process with a better chemical control of the converter practice. To increase steel production and reduce fuel and labour costs, the E.T. management decided to by-pass cupola furnaces and transfer molten pig iron directly from the blast furnace to the converter. Consequently, the cupola furnaces for melting pig iron and scrap were no longer used at the E.T. Works after 1882. With this practice, however, difficulties were experienced in the smooth operation of the converter because of unpredictable fluctuations in the silicon and carbon contents of the blast furnace product. The need to circumvent this operational problem prompted Captain Jones to the idea of a metal mixer to even out by dilution the composition variations of hot metal from one blast furnace tap to the next. Incidentally, Jones' invention of a metal mixer became the subject of patent litigation for many years. It was in 1887 that Captain Jones installed his first metal mixer of 100 ton capacity at E.T. With a successive series of innovations, notably by Holley and Jones, the Edgar Thomson Works became a world-class prototype for the mass production of Bessemer steel rails.

It would be appropriate to mention here briefly the events leading to the formation of the Carnegie Steel Company and ultimately to the United States Steel Corporation. To expand steel production and also fend off competition, Carnegie acquired the Homestead Steel Works in 1883, Duquesne Steel Company in 1889 and in the following year two more steel works and the controlling interest in six others. In 1899 all Carnegie's holdings in the steel mills, coal mines, coke ovens, iron ore mines, railroads etc., were brought together as the Carnegie Steel Company.

A very successful entrepreneur in promoting the growth of the steel industry in the United

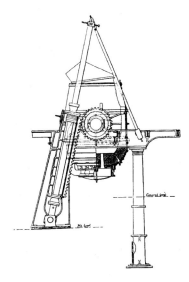

Fig. 2   Edgar Thomson Bessemer converters with Holley detatchable bottoms. From Ref. 13.

States, Andrew Carnegie expressed his admiration of and gratitude to Sir Henry Bessemer by stating in one of his writings:[14]

> '*Rest easy, 'Great King of Steel,' and smile on all attempts to rob you of immortality and of the gratitude of the world.*'

In 1901, Charles Schwab (then the President of the Carnegie Steel Company), Elbert H. Gary (the President of the Federal Steel Company) and J.P. Morgan (the most influential international banker) worked out a 'Steel Deal' to buy the Carnegie Steel Company* and combine it with the Federal Steel Company to form the new United States Steel Corporation. For additional information reference may be made to a recent paper by the author on the history of USX Corporation (*Ironmaking Steelmaking*, 1996, **23**(4), 289–292).

STEEL RAIL PRODUCTION FEVER IN THE UNITED STATES

As noted from the data in Fig. 3, the impetus to the growth of the steel industry in the United States, that started in the last quarter of the nineteenth Century, was the extensive transcontinental railroad construction. The highest peak of annual rail production (3 977 887 tonne) was in 1906. However, with respect to the total steel production (Bessemer and open hearth processes), the highest percentage of steel produced was used in making rails during the period 1875–1885, as depicted in Fig. 4.

The quantities of Bessemer steel and open-hearth steel used in rail making are compared

*In this transaction, Carnegie's price tag was $303 450 000 in 5 percent bonds and $188 556 160 in common and preferred stocks of the new Corporation, a total of $492 006 160; at that period the average steelworker earned $500 a year.

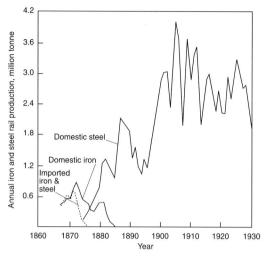

Fig. 3   Annual rail production in the United States. Data from Ref. 16.

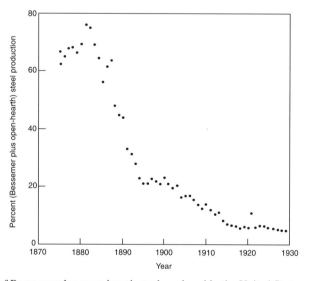

Fig. 4   Percent of Bessemer plus open-hearth steel produced in the United States used in rail production. Data from Ref. 16.

in Fig. 5. From 1907 to 1909 there was a dramatic drop in the use of Bessemer steel for rail making. After 1910, rails were made primarily from open-hearth steel.

For a comprehensive review of the growth of the railroad construction in the United States, initially inspired by the Bessemer steelmaking technology, reference should be made to a recent book by Misa,[7] entitled '*A Nation of Steel – The Making of Modern America 1865–1925.*'

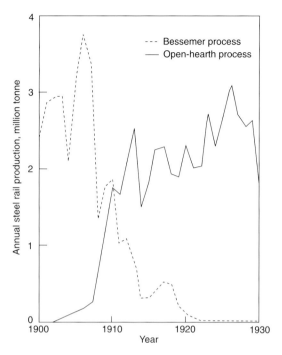

Fig. 5   Transition from Bessemer steel to open-hearth steel in rail production in the United States. Data from Ref. 16.

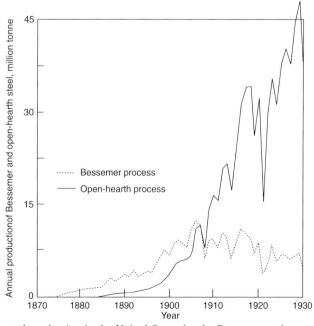

Fig. 6   Annual steel production in the United States by the Bessemer and open-hearth processes. Data from Ref. 16.

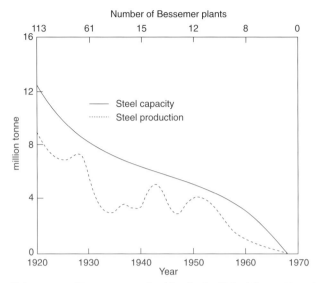

Fig. 7   Declining years of Bessemer steelmaking in the United States. Data from Ref. 17

OPEN-HEARTH PROCESS REPLACING BESSEMER STEELMAKING

The increase in steel production in the United States by the Bessemer and open-hearth processes over a period of 55 years is shown in Fig. 6. After 1907, there was a rapid rise in steel production by the open-hearth process, while the annual production by the Bessemer process remained at a moderate level of about 7.5±1.0 million tonne. Decline in steel production by the Bessemer process after 1920 is shown in Fig. 7. The number of Bessemer plants was reduced from 113 in 1920 to 8 in 1960. This mode of steelmaking ceased to operate in the mid 1960s.

Because of the abundance of low phosphorus iron ore deposits in the Lake Superior Iron Range, only acid Bessemer steelmaking was practiced in the United States. Companies which had no access to iron-rich, low phosphorus ores, which were locked out by the Carnegie Steel Company in the 1890s then by the U.S. Steel Corporation after 1901, had to adopt open-hearth furnaces for steelmaking by the duplex practice.

Originally, American acid Bessemer/basic open-hearth duplexing was developed by Benjamin Talbot, an English metallurgical chemist who was recruited by the Southern Iron Company in 1889 to install open-hearth furnaces at its mill in Chattanooga, Tennessee. Ultimately, Talbot designed a 75-ton tilting open-hearth furnace, which could be charged with partially converted Bessemer metal for final refining in the basic open-hearth furnace to the desired steel grade.

In the 1940s and thereafter, most of the Bessemer steelmaking was replaced by acid Bessemer/basic open-hearth duplexing, known as 'liquid-metal practice.' After fettling the furnace lining, scrap and limestone were charged, followed by 'soft' blown converter metal and later by hot metal from the mixer in appropriate proportions to give at clear-melting a carbon content of 0.5 percent above the aimed carbon content at furnace tapping. The Bessemer converter was also used to a limited extent with the electric-arc furnace.

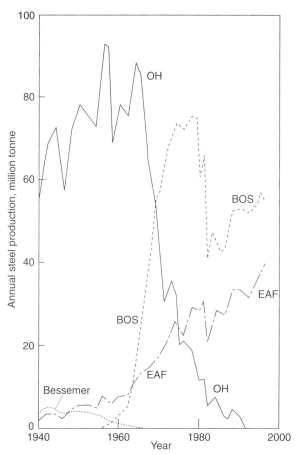

Fig. 8 Changes in steelmaking processes in the United States after 1940. Data from Ref. 17.

CHANGES IN STEEKMAKING PROCESSES AFTER 1940

Annual steel production data for the United States given in Fig. 8 illustrate the dramatic changes that occurred in steelmaking processes during the past sixty years. Between 1964, and 1971, there was a sharp drop in basic open-hearth steelmaking (OH) and an accompanying sharp rise in both top and bottom blown basic oxygen (BOS) and electric-arc furnace (EAF) steelmaking. Between 1974 and 1979 the average annual steel production was in million tonnes: 74 by BOS, 28 by EAF and 18 by OH.

For a variety of reasons, there was a noticeable decrease in steel production in the United States in the early 1980s. This decline in steel production was due mainly to;

(i)   closures of old and obsolete steel plants,
(ii) modernising BOS (BOF, BOP & Q-BOP) and continuous casting plants,
(iii) installation of new plant facilities,
(iv) adjustment of production to domestic and export needs.

By the mid 1980s, the steel companies consolidated their various steelmaking and raw materials subsidiaries and divisions through a series of reorganisations and restructuring. In the mid 1980s, the average annual steel production was 55±2 million tonne by BOS and 36±4 million tonne by EAF. All open-hearth steelmaking operations ceased after 1990.

SUMMARY OF 140 YEARS OF EVOLUTION IN STEELMAKING TECHNOLOGY

It is appropriate to present here briefly, in a graphical form, an overview of the 140 years of evolution in the steelmaking technology. The data in Fig. 9 up to the year of 1979 are taken from the 1979 Metals Society Presidential Address by Van Stein Callenfels,[18] the statistical data for the period 1979 to 1996 are from Ref. 17.

The basic Bessemer (Thomas) process was started in the early 1880s; both in Great Britain and Europe. After 1905, the basic Bessemer steelmaking exceeded the acid process, because of the more readily available high-phosphorus iron ores which were cheaper than the low- phosphorus ores. In the year 1907, the world steel production by acid and mostly basic open-hearth processes equalled the amount produced by the acid and basic Bessemer processes. After 1907, the open-hearth steelmaking in all the industrial countries over shadowed the converter steelmaking operations.

Exponential growth of the world crude steel production appears to have ended in 1979. Since then, there has been a cyclic up and down in the world steel production at about five year intervals, similar to the trends in the United States. For comparison purposes, the steel production in the United States relative to the total world production is shown in Fig. 10. The peak years for steel production in the United States, relative to other industrial countries, were during and subsequent to the times of World War I and II.

The advent of large capacity top and bottom blown oxygen steelmaking (BOS), EAF steelmaking, continuous casting and automation of plant operations, resulted in the reduction of the number of people employed (wage and salary earning) in the steel industry as a whole. While the world steel production (excluding P.R. China) increased slightly from 620 million tonne in 1975 to 660 million tonne in 1995, the total number of people employed by the steel industry decreased from about 2.5 million in 1975 to 1.35 million in 1995. The resulting increase in the average annual steel production per employed person from 1975 to the present time is shown in Fig. 11.

In this statistical analysis, the Chinese steel industry is not included, because the number of people they employ is about three times that in the rest of the world steel industry. For example, the steel production per person employed in P.R. China has been 21 tonne per person between 1990 and 1995, as compared to the average world production of 400 tonne per person in 1990 and 500 tonne per person in 1995.

The statistical data in Fig. 9 for the past fifteen years would suggest that for the next decade or two, steelmaking will be almost entirely by the BOS (using blast furnace hot metal) and EAF processes and the total annual world production fluctuating presumably between 750 and 850 million tonne.

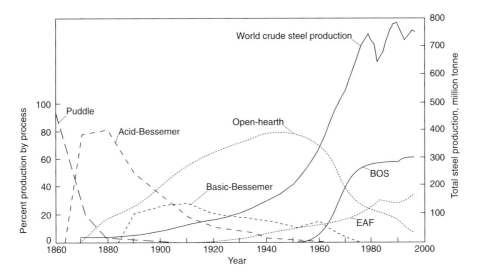

Fig. 9 Development of major steelmaking processes. Data from Refs 17 and 18.

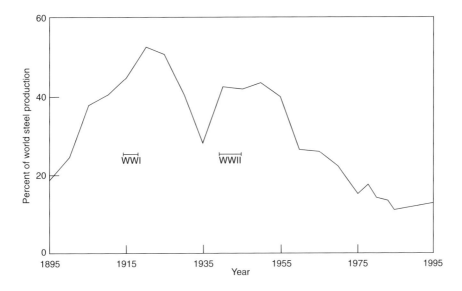

Fig. 10 Steel production in the United States relative to the total world steel production. Data from Refs 16 and 17.

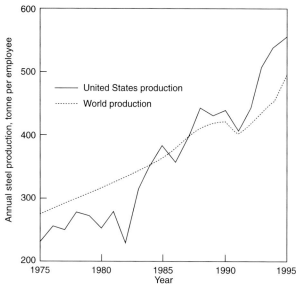

Fig. 11   Changes in average steel production per employee in the United States and in all other countries with exception of P.R. China. Data from Ref. 17.

CONCLUDING REMARKS

On retrospect, one might surmise that evolutions in the steelmaking technology during the past century were, to some extent, the direct and indirect consequences of the Bessemer process. At the time of Bessemer's inception of the basic concept of his process, very little was known about the chemistry of steelmaking reactions. Solutions of initial technical difficulties in the development of the Bessemer process and subsequent demands by the steel users for Bessemer steel of improved metallurgical quality opened up opportunities for process developments by the metallurgical chemists and engineers.

Bessemer's early recognition of the converter steel being more malleable for hot working when made using pig iron with high manganese and low sulphur contents, enlightened Robert Mushet to the idea of adding spiegeleisen (an alloy of iron containing 15 to 30% Mn and 4 to 5% C) to the blown steel in the ladle.[19] Noting the importance of manganese, Bessemer then developed his own ferromanganese made with a much higher Mn/C ratio than in spiegeleisen, so that the ferromanganese-treated low-carbon steel could be made as required for shipbuilding and other applications. Evidently, the concepts of deoxidation and the necessity of high Mn/S ratio in the steel to suppress hot-shortness were inadvertently borne out from the first use of ferromanganese to treat converter steel in the ladle. However, the technical significance of these metallurgical events might not have been understood at the time.

Another metallurgical innovation was that brought about by Sydney Gilchrist Thomas. His experimental work at the steel works in Cleveland (England) in the late 1870s revealed that by lining the converter with blocks of fritted calcined dolomite and using a lime-rich slag, the steel could be dephosphorised during the converter blow. Soon after the discovery by this metallurgical chemist, the acid Bessemer practice in Great Britain was changed to what then became the basic Bessemer (Thomas) process. Similar changes occurred in the European steel works by the mid 1880s. Also, the upcoming open-hearth process readily adopted the basic steelmaking process of Thomas. For a detailed disclosure on how the concept of basic steelmaking started, with the object of steel dephosphorisation, reference may be made to a paper by Almond.[20]

It is worthy of note that during his visit to the steel plants in the United States, Thomas sold his process rights to the Bessemer Association, through Carnegie, in May 1881 for $275 000. For this business transaction, Carnegie took a very handsome sum of $50 000 as commission. Because the Bessemer Association was in sole control of the Mesabi iron ores of low-phosphorus content, they intentionally withheld Thomas' license to fend off competition from the Southern States where iron ore deposits are high in phosphorus. In fact, it was for this reason that the basic Bessemer (Thomas) process was never practiced in the United States.

The foregoing reminiscences are just a few examples of innovations in the steel industry in the late nineteenth Century, which were inspired by the onset of the Bessemer process. By the mid twentieth Century, when Bessemer steelmaking (acid and basic) was being replaced, to a large extent by the basic open-hearth process and the upcoming EAF process, there emerged a new process of steelmaking with oxygen top blowing in the converter. The initial work in the development of this process was done at Linz-Donawitz (Austria) in the mid 1950s; which happens to coincide with the Centenary of Bessemer's first announcement of his process at the Cheltenham Meeting of the British Association for the Advancement of Science. Emergence of the Linz-Donawitz (LD) process may indeed be looked upon as the re-birth of the Bessemer process, using pure oxygen instead of an air blast. In fact, the present basic oxygen steelmaking processes, e.g. LD (BOS or BOF) and OBM (Q-BOP), may well be considered rightfully as the ultra-modern hybrids of the Bessemer process, with a larger converter capacity, sophisticated equipment and measuring devices for computer assisted, or even fully automated steelmaking operations.

REFERENCES

1. W.F. Durfee: *Amer. Soc. Mech. Engs Trans.*, 1885, **6**, 40.
2. A.L. Holley: *A Treatise on Ordnance and Armor*, D. Van Nostrand, New York, 1865.
3. R.W. Hunt: *Amer. Soc. Mech. Eng. Trans.*, 1885, **6**, 62.
4. A.L. Holley: *Engineering* (London), 27 March 1896, **6**, 414.
5. J.R. Stubbles: *The Original Steelmakers*, The Iron and Steel Society Inc., Warrendale, 1984.
6. K.C. Barraclough: *Steelmaking 1850–1900*, The Institute of Materials, London, 1990.
7. T.J. Misa: *A Nation of Steel – The Making of Modern America 1865–1925*, The John Hopkins University Press, Baltimore, 1995.
8. A.L. Holley: *Journal of the Franklin Institute*, 1872, **94**, 252.
9. H.M. Howe: *Engineering and Mining Journal*, 1875, **20**, 258; *Trans. AIME*, 1876, **5**, 515.
10. A.L. Holley: *Trans. AIME*, 1875, **4**, 138.

11. H.M. Howe: *The Metallurgy of Steel, 2nd rev. ed.*, Scientific Publishing, New York, 1891.

12. On Carnegie's career see Pennsylvania Railroad General Order 10, 21 November 1859, vol. l, ACLC; letters to Enock Lewis in vols 1–3, ACLC; and the 'farewell address,' Andrew Carnegie to Pittsburgh Division PRR, 28 March 1865, vol. 3 ACLC. (ACLC an abbreviation of 'Andrew Carnegie Papers, Manuscript Division, Library of Congress, Washington, D.C.

13. A.L. Holley and L. Smith: *Engineering* (London), 19 April 1878, **25**, 296.

14. B.J. Hendrick: *The Life of Andrew Carnegie*, Heinemann Publ., London 1933, 145.

15. E.T. Turkdogan: *Ironmaking & Steelmaking*, 1996, **23**(4), 289.

16. Annual Statistical Report of the American Iron and Steel Institute for 1912 to 1930.

17. Annual Statistical Report of the American Iron and Steel Institute for 1920 to 1996.

18. G.W. van Stein Callenfels: *Ironmaking & Steelmaking*, 1979, **6**, 145.

19. R. Mushet: *UK Patent No. 2218*, 22nd September, 1856.

20. J.K. Almond: *Ironmaking & Steelmaking*, 1981, **8**, 1.

The Development of the Bessemer Process in Continental Europe

F. TOUSSAINT

THE BEGINNING OF THE BESSEMER PROCESSES ON THE CONTINENT AND ECONOMIC CONSEQUENCES

After the invention of the pneumatic process by Henry Bessemer between 1855 and 1860, the first to apply the process in Germany was Krupp in Essen in 1862, (first melt blown on 16 May 1862). Krupp was informed about the development by Bessemer in England, because his representative agent in Britain, Richard Longsdon, was a brother of Bessemer's friend and collaborator Frederick Longsdon. So he applied for a patent according to Bessemer's indications, but this was rejected by the German Imperial Patent Office, which generally at that time was not very much in favour of inventors. Since Bessemer had no patent in Prussia, Bessemer and Krupp were interested in keeping the fact of the existence of tiltable converters in Essen a secret. The name of the steel plant in Essen was disguised and called 'Räderwerk C' (wheelshop C).[1] By 1867 Krupp was operating 18 converters, thus being the most important Bessemer steelworks on the continent. The second Bessemer plant in Prussia was erected at Hörde in 1864 (where later the first Thomas heat was blown). It seems that Krupp was the first to arrange three vessels in a row, instead of two in a circular arrangement (Fig. 1).

The Prussian Government nevertheless must have been interested in the process, because already in September 1856 tests were being made in the state-owned Königshütte in Silesia, but those were not successful. Königshütte definitely started Bessemer steel production in 1864. This proves again that the Prussian government was interested in promoting industrial progress and gave with her state plants an example for private enterprise, in spite of setbacks.[2] By 1877, that is just before the basic process came to the continent, within the German Empire altogether 76 Bessemer converters were in operation, most of them (64) in Prussia, and 54 of these, that is more than 70%, were in the Ruhr area.

In France, according to Gille and Cournot, the first Bessemer plant was started in 1858 at Saint-Seurin, near Bordeaux with three converters, thus being the very first producing Bessemer plant on the continent.[3] The first vessel contained 1.5 t of steel, the two later ones held 3.0 t. The next plant with two vessels was at Imphy in central France. The converters here came from Saint Seurin in 1862, because the owner, James Jackson, went bankrupt.[4] Le Creusot followed in 1863 until, by 1869, 17 vessels were in operation.[5]

In France important research had been conducted in the second half of the fifties and the early sixties, which came very close to the invention of Bessemer. Caly-Cazalt obtained patents in France which were really rather similar to Bessemer's ideas.[6]

Fig. 1   Bessemer steelworks of Krupp in Essen. About 1910.

Also in 1863, like Le Creusot, the famous Austrian metallurgist Peter Tunner[7] (who died one year before Henry Bessemer) blew the first melt in Austria, at Turrach, Styria, on the 19 November 1863. In 1864 Bochumer Verein in the Ruhr area and Königshütte in Silesia started the process in their plants. In the same year the process came to the United States. Tunner predicted, correctly, that the wrought iron production would come to an end against the competition of better and cheaper ingot steel produced by the Thomas process (Fig. 2).[8] In Austria after Turrach the next plants were at Heft in Carinthia and Graz in Styria.

The average yearly increase in 10 years from 1870 to 1880 in the production of acid Bessemer steel is shown in Table 1. Most countries increased in this period their production by around 17% per annum with the exception of Belgium and specially the US, were the production exploded and even surpassed the production of Great Britain in 1879. In 1880 it was more than a million tons.

Table 1   Average increase of Bessemer steel production from 1870 to 1880

	UK	Germany	France	Austria	Belgium	Sweden	USA
Ø per year 1870–80	19.5%	18.6%	16.6%	13.8%	31.8%	17.5%	39.9%
Source: Bernhard Osann: Lehrbuch der Eisenhüttenkunde, Leipzig 1926							

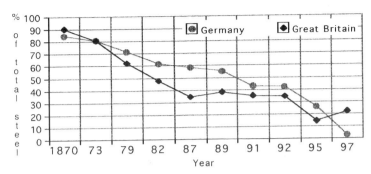

Fig. 2 Decrease of wrought iron production

The introduction of the Bessemer Process on the continent resulted in a search for low phosphorous ores. For example in France, where the Algerian ores became very interesting, the iron and steel plants began to move towards the coast to get better transport conditions for such raw material.[9] In Germany especially, Spanish ore deposits were developed and Krupp in 1871/2 bought mines in northern Spain (see Table 2). In Britain this was not the case to such an extent, because it was sufficiently equipped with suitable ores, and the US had huge quantities of them.

Table 2 Bessemer steel (acid) production in Germany

	Krupp	Bochumer Verein	Hörder Verein	Poensgen	Königshütte	Total
1862	423			1 366		
1865	17 905	732	2 645	2 250	43	
1870	61 596	19 888	11 614	?	0	125 000
1875	69 355	54 135	?	?	7 112	241 000
1879	124 051	65 162	?		22 379	465 000
1880			Thomas			686 000
Source: UlrichTroitzsch: Die Einführung des Bessemer Verfahrens in Preußen – ein Innovationsprozeß aus den 60er Jahren des 19, Jahrhunderts, Frank R. Pfetsch, Innovationsforschung als multidisziplinäre Aufgabe, Göttingen 1975						

Thus the competitiveness of continental plants became worse and worse compared to Britain, where the cheap Bessemer process was based on excellent ore deposits close to the steel plants.

The puddling process allowed the processing of at least medium phosphorous ores with a satisfying result, but its productivity remained low (Fig. 3), whereas the productivity of the pneumatic processes were extremely high as compared to wrought iron production. So the inability of the acid Bessemer Process to treat phosphorous ores was felt very strongly on the

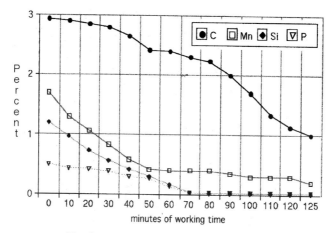

Fig. 3   Chemistry of the puddling process

continent, where the high productivity of the Bessemer process could only be used to a limited extent.

Already John Percy (though a little bit satirical on Bessemer's process: 'what a magnificent pyrotechnic effect') said in 1864[10]

'In order that the Bessemer process should be generally applicable in this country, it must be supplemented by the discovery of a method of producing pig iron sensibly free from sulphur and phosphorus with the fuel and ores which are now so extensively employed in our blast furnaces. The problem may be difficult of solution, but surely it is not a hopeless one'

Indeed it was not a hopeless one, but nevertheless it took another 15 years to solve the problem. When Thomas solved it, an explosion of production started on the continent similar to that which occurred in the US with the acid Bessemer process. Now the enormous deposits of phosphoric ore in the area called today Sar-Lor Lux, called minette, (an oolithic ore) became valuable, Lorraine became an important iron and steel producing area, the Sarre and Luxemburg equally. But other locations also became interesting like Peine[11] and the Upper Palatinate in Germany, Bohemia and Moravia in Austria. Later the north Swedish iron ores from Kiruna and Gellivara became a main source for the continental steel industry, thus enabling Germany to surpass Britain in steel production by the end of the century . Though the transformation cost of the Thomas process was somewhat higher than the acid process, the pig iron was much cheaper, so that the Thomas steel was cheaper than Bessemer steel and also cheaper than puddled steel.

In an American Report of 1889[12] it was stated that: '.... the United States and Germany have made more progress than any other countries, and very much more relatively than Great Britain.' The reason was mainly in the United States the introduction of the Bessemer process (acid) and in Germany of the Thomas process (basic), both British inventions.

An old economic problem was strongly influenced by from the Thomas process: the canalisation of the Moselle. In the second half of the 19th century there was no one iron producing district left which possessed both ore and fuel resources, so transport facilities and cost had an increasing influence on iron production. The Ruhr area had huge fuel resources but no ore deposits of economic importance and most of the ore had to be imported from abroad, mainly from Spain and Sweden. A certain amount could be obtained from the Siegerland (Mn–rich, but costly ores). The introduction of the Thomas process changed the situation dramatically. Now the Ruhr plants were very keen to get the Lorraine minette as cheaply as possible, thus favouring the construction of the Moselle canal. The rich deposits of Lorraine were the only chance for the German steel industry (Ruhr) to start the modern mass production of iron and steel, which the former ore deposits and manufacturing methods did not allow. But the Ruhr works were the only ones in Germany to favour this project. All the others were strongly opposed, because it hurt their own interests. So in 1890 the Prussian Government convoked a meeting in Coblence with all interested parties. The question the Government put to the assembly was:

> 'How far the development of the basic Bessemer process is of importance for the competitiveness and the maintenance of the export trade of the Rhinish–Westphalian iron industry and the German iron industry in general?'

Thus the question of the Thomas process and the consequent substitution of wrought iron by ingot steel became fundamental to the decision on the Moselle canalisation, By the turn of the century, when the north Swedish ores with high phosphorus content and high Fe content became available, the Moselle canal lost its significance for the Ruhr industry and construction of the canal was finally decided in more recent times, when these specific questions were no longer relevant.[13]

The question of the economics of the process was very much discussed, also controversially.[14] But within a short period the process became accepted in all European countries which had phosphorus ore deposits. The technical activity and minor and major improvements reduced the cost, thus decreasing the difference between the production cost of the acid and basic processes. Additionally the sale of Thomas slag for agricultural use added considerably to the economy of the process, thus making less important the difference between the cost of Bessemer and Thomas pig iron.

The growth of the Thomas steel production in the different countries is shown in Figs 4 and 5. This shows clearly the extraordinary importance of the Thomas process for the German economic development. By 1892 Germany produced more than 2 million tons of Thomas steel, which was about two thirds of the world production. The anual growth of Thomas steel in Germany between 1883 and 1909 was an average of nearly 13%, an absolutely extraordinary figure. In 1909 the production of Thomas steel in Germany exceeded the total steel output of Britain (Fig. 6).

INTRODUCTION OF THE THOMAS PROCESS TO GERMANY

It is well known today that Sidney Gilchrist Thomas tried to propose his ideas at the spring meeting of the Iron and Steel Institute in London in March 1878 during the discussion of the paper by Lowthian Bell on his procedure to eliminate phosphorus from pig iron by washing

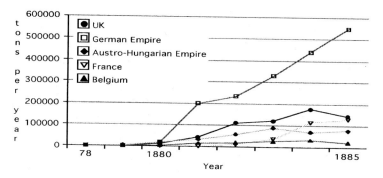

Fig. 4   Basic Bessemer steel production up to 1885.

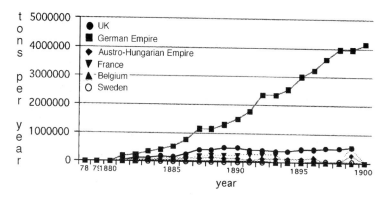

Fig. 5   Basic Bessemer steel production up to 1890.

it with iron oxides at low temperatures. The assembly 'simply took no notice of his unde-monstrative announcement.'[15] At the autumn assembly, which took place in Paris, Thomas submitted a paper on his process, which was placed at the end of the agenda and finally not read 'due to lack of time.' The fact of the non-publication of his paper turned out to be advantageous, because 'there would have been ample and unlooked for compensation.'[16] At Paris he met Mr Richards, manager of the Bolckow, Vaughan and Co. works in Middles-brough during a visit to the Creusot works. This was for him the opportunity to get this important steel works situated next to the main English deposit of phosphorus iron ore as a partner, who was able and willing to develop his process. On the 4th of April Thomas and Richards made a couple of experimental melts producing good rails from Cleveland iron, which became known immediately in the technical press. 'The news spread rapidly and wide, and Middlesbrough was soon besieged by the combined forces of Belgium, France, Prussia, Austria and America.'[17]

Thomas applied for patents in the German Empire on the 26th of March 1878, which apparently did not attract the attention of German steel manufacturers, and another applica-tion was made on the 10th of April of 1879. With the help of Hörde and Ruhrort, after the signature of the contract between Hörde and Ruhrort on one side and Thomas on the other on the 26th of April, difficult negotiations took place at the Imperial patent office in May and November 1879, which ended finally in the rejection of the oppositions and the patent was

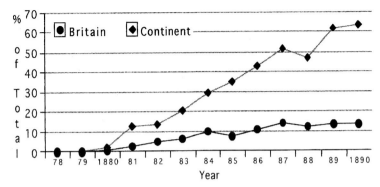

Fig. 6 Proportion of basic steel.

definitely granted in November 1880 or in January 1881, more than a year after the first successful melts in Ruhrort and Hörde. All details of the complicated procedure of the patent application in Germany, which suffered opposition from most of the German steel producers, including Krupp who was oriented towards acid Bessemer with his ore supply policy, can be found in a summary in the Thyssen archives.[18]

Mr Josef Massenez, president of Hörder Bergwerks- und Hüttengesellschaft in Hörde near Dortmund, asked his steel works manager Richard Pink (an Englishman who worked for Hörde after having erected the Bessemer plant there) to send immediately a cable to Mr Richards of Bolckow, Vaughan and Co. (a friend of Massenez) to ask for permission to inspect the new process.[19]

On arrival of the positive answer Massenez, his colleague Meier and Pink travelled immediately to Middlesbrough, where they met Mr Pastor together with Dr Graß from Rheinische Stahlwerke at Ruhrort (near Duisburg) who had arrived before them. On the 26th of April 1879, only three weeks after the first public demonstration, both companies signed a contract with Thomas for Germany and Luxemburg with the exception of the Company of Aug. Metz & Co.[20] in Luxemburg. Emile Metz and Jean Meyer, representatives of Aug. Metz & Co. had signed just a few days before, on the 21st April[21] an agreement with Thomas, but only for their own company, thus leaving freedom to Thomas for other agreements. Strangely enough, production was started by Metz in their plant at Düdelingen (Dudelange) only on the 15th of April 1886. The reason for this might be the fact that Düdelingen was not connected to the railway system until 1884.[22] The plant at Düdelingen (Düdelinger Actienverein) which was founded by Aug. Metz & Co. and SA. des mines du Luxembourg et des Forges de Sarrebruck (all companies which later formed the ARBED group (Acieries réunies de Burbach, Eich et Dudelange) inherited the Thomas contract from Metz. Thus Metz was not the first producer, but certainly the first licensee of Thomas if one excludes the informal agreement with Kladno, Bohemia (see later). The Düdelingen steelworks very soon became a progressive plant, as is shown by a paper published in 1908.[23]

This contract with the German firms, which included a fixed sum plus a tonnage fee, was superseded one year later, on the 12th of April 1880 by a new contract replacing the old one by a monetary compensation of £26 400 (equal to about 1/2 million Reichsmark, payable in three equal yearly rates of £8500 plus an interest rate of 5% per annnum. This sum was about six times greater than Bessemer could get out of Germany altogether (though he received millions of pounds from other countries, were he did obtain patents).[24]

On the 8th of May 1879 Thomas read a paper in London at the Iron and Steel Institute before a large audience. In this paper he mentioned the works of Professor Gruner in France, who already in 1867 was aware of the fact that the acid lining and slag was detrimental to dephosphorisation. A few days after the meeting, on the 13th of May, Massenez travelled to Eston[25] were Richards showed him the process in an 8 tons vessel. In the evening of the same day, he met Thomas again and concluded an agreement with him on the rights for the Austro-Hungarian Empire. Massenez had many friends in Austria, since he had studied at the mining academy of Leoben with the famous Professor Tunner with whom he became very good friends.

According to Ludwig Beck,[26] a few days after the visit to the Middlesbrough plant , on the 10th of April, a trial melt was blown at Hörde with a phosphorite lining, a method which was soon abandoned since Massenez became convinced that dead burnt dolomite was the most appropriate material. Although the author could not find any evidence for further melting at Hörde between the 10th of April and the official inauguration date of the 22nd of September 1879, there must have been an intensive test campaign at both Hörde and Ruhrort, which was not difficult to accomplish since both plants had Bessemer converters, (Ruhrort six and Hörde five vessels).[27] This is important to state, because of the priority discussion between Bohemia and the Ruhr area. According to the facts found in the archives of Hoesch in Dortmund and in Tunner's report, it appears probable that the first heat blown on the continent was in Kladno, Bohemia in March 1879 at the request of Thomas himself (see Austria), but on a production scale the first was certainly in the Ruhr area in September 1879.

Several attempts had been made to solve the problem of phosphorus in pig iron, in Bohemia (see later sections), in France and also in Germany, where two Krupp engineers (Bender and Narjes) applied for a patent.[28] In contrast to most of the German steel producers, Alfred Krupp was not in favour of the Thomas process, because it did not fit to his company's policy, since he was engaged in steelmaking from Spanish low phosphorus iron ores. Consequently he tried to oppose the Thomas patents in Germany, but without success. These quarrels were the reason why the Thomas patent in Germany was finally issued in January 1881, more than a year after the successful start of the process in Hörde and Ruhrort.

On the same day, the 22nd of September 1879, the first melts of Thomas steel, as it was called all over continental Europe from the very beginning, were successfully blown in Ruhrort and in Hörde (Figs 7, 8 and 9).

Though the basic facts for a successful execution of the process were identified in the licence agreement as:

(1) dolomite lining baked with dehydrated tar,
(2) limestone addition to the melt after the first blowing period and
(3) the so-called afterblow (continuation of the blowing after the end of the decarburisation).

for the complete elimination of phosphorus, a lot of experience had to be gained. Both the Hörde and the Ruhrort plants did a lot of research and experimental work to improve the process and make it more economic.

One important point was the manufacture of the basic refractory lining for the converters.

Fig. 7   Thomas steelworks in Duisburg-Ruhrort, before 1906.

Fig. 8   Thomas steelworks in Duisburg-Ruhrort.

Fig. 9   Perhaps the first photograph of a mixed Bessemer/Thomas steelworks. Rheinische Stahlwerke, Meiderich, shortly after the first Thomas melt in 1879.

Mr Pastor, the general manager of the Ruhrort works, was well acquainted with Dr Vygen, the owner of a refractory manufacturing plant in Duisburg. He succeeded first in producing bricks of high quality. So Ruhrort and Hörde signed a contract on the 4th of June 1879 with Vygen and two other manufacturers, Dr C. Otto in Dahlhausen (the world renowned coke oven plant manufacturer) and Stolberger AG für feuerfeste Produkte.[29]

At the General Assembly of the Technischer Verein für Eisenhüttenwesen (the precursor of the VDEh) on the 14th of December 1879 in Düsseldorf Massenez reported about the progress of the Thomas process in Germany. It was impressive to hear about the enormous progress achieved in the two plants, since both companies had been in production for less than 3 months. Massenez reported that until recently their specialists were convinced that a minimum silicon content in the pig iron was required to achieve the necessary temperature in the steel. In Eston up to 3% silicon content was accepted in the pig. But it was found that above 1.7% Si severe trouble appeared with the durability of the converter lining. To solve those problems in Bohemia a two stage converter process (acid + basic), was used for some time. The work carried out at Hörde showed that lower silicon would be preferable plus higher P, and silicon could be easily replaced by phosphorus to achieve hot melts which could be cast without problems by bottom pouring (Fig. 10).

At Hörde the durability of the lining was up to more than 100 heats. Massenez was not yet satisfied with the life of the bottoms, which was not higher than 15 heats, but he was confident that these figures could be improved. He was right: in 1884 Professor Wedding esti-

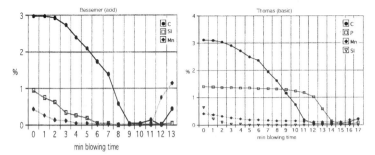

Fig. 10 Chemistry of the acid and basic Bessemer processes.

mated at that time the average life for the lining of a converter at 100 to 120 heats and for the bottoms 18 heats.[30] The system designed by the American Holley (interchangeable bottoms) for the acid converter and generalised in Bessemer steelworks, was adapted to the Thomas process from the very beginning. In 1909, 30 years later, the durability of the converter bottoms achieved 64 heats and the lining 280 heats.[31] In the last years of operation of the process the average bottom life was 50 to 90 heats, the life of the lining 300 to 400 and in some cases up to 600 heats in converters which were much larger than those used around 1900 (60t instead of l0t), (Fig. 11).[32]

Massenez said that he was able to treat without problem ordinary white pig iron containing less than 0.5% Si, about 2.5% C and more than 2% P. He explained that (at least under German conditions) the price difference for silicon-rich Bessemer pig iron and ordinary white iron was about 30RM (at that time £1.50), which makes clear the extraordinary economic importance of the Thomas process. Finally Massenez spoke about *your enemy, our friend the phosphorus*, because it was a cheap and very effective source of heat (at that time he did not yet know about the utility of the slag as a fertiliser). So the new experience was, that silicon could be replaced by phosphorus (see later: Joseph Gängl von Ehrenwerth), and that the transformation cost was not much higher than with the acid Bessemer process. Massenez estimated the extra cost for lime at 1.50 to 1.80RM (= 1s 6d to 1s 10d) per ingot ton, and the extra cost for refractory material at 1.00 to 1.50RM (= 1s to 1s 10d), or together 2s 6d to 3s 8d, to which the licence fee must be added. The difference for the pig iron freight paid to Dortmund actually was 27RM (= £1–7s) He showed several analyses and results of measurement of tensile strength and reduction of area, proving that the quality of the steel was excellent.[33] So in just three months of practical operation very important progress was made, which made in a very short time the Thomas process an extremely economical and technically feasible method of steel making.

The managing directors of Rheinische Stahlwerke at Ruhrort (today Duisburg), Mr Pastor (of Belgian descent) and of Hörder Berg- und Hüttenverein at Hörde (today Dortmund), Mr Massenez, seem to have been not only excellent engineers (surrounded by equally good metallurgists), but also very efficient businessmen. The number of international contracts they concluded in the course of a year is impressive, especially when one looks at the economic consequences. One of the metallurgists who participated in the success of the process was without doubt the already mentioned Englishman Pink.[34] He left Hörde suddenly, to-

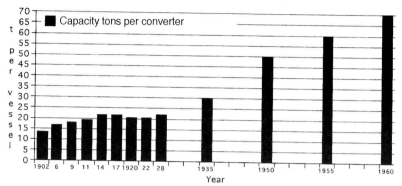

Fig. 11   Development of the size of converter vessels in Germany.

gether with his German colleague, when he learned that his superior Massenez was awarded the important sum of 80 000RM (equal to £4000), whilst both engineers were left in the cold.[35]

The patent litigation at the Imperial Patent Office by Thomas was not so much a quarrel between Thomas and the office as between Krupp and the other German steel manufacturers, which Krupp lost and therewith his leadership in mass ingot steel making in Germany (though certainly not loosing his leadership in special steels).

To be in possession of all possibilities both companies bought the patent rights granted to Jean Marie Harmet from Lyon, France on the 27nd of February[36] (valid as of 20th March 1880) which proposed the separation into two processes (combustion of silicon in a converter with an acid lining and of phosphorus in a vessel with a basic lining) for pig iron with high silicon and low phosphorus content. This process was applied for some time at Witkowitz. Such pig iron in the first stage produced a large volume of silicon rich slag, which was not only dangerous for the basic lining but also hindered the solution of phosphorus in this kind of slag (the old disadvantage of the acid Bessemer process).

The next licensees in Germany were:

1879   de Wendel in Hayingen, Lorraine (today Hayange,France) (June)
       Stumm, Neunkirchen, Sarre(June)
       Dietrich, Niederbronn (July)
       Lothringer Eisenwerke, Ars (Moselle) (November)
       On the 11th of November 1879 a written request for a licence was sent to
       Hörde by Dillinger Huttenwerke (Sarre)

1880   Rothe Erde, Aix-la-Chapelle (Rhineland) (January)
       Bochumer Verein, Bochum (Westphalia) (March)
       Maximilianshütte, Rosenberg (Bavaria) (April)
       Burbacher Hütte, Sarre (January)
       von Gienanth, Kaiserslautern (Palatinate)

1881 Gutehoffnungshütte, Oberhausen (Rhineland)
      Peiner Walzwerk, Peine (Lower Saxony)[37]
      Dortmunder Union, Dortmund (Westphalia) (March)
      Phoenix, Laar (Rhineland) (April)[38]

1883 Hoesch, Dortmund (November)
      Stahlwerk Krieger (December)
      In Upper Silesia, one of the most important areas of mining and metallurgy in
      Prussia,
      Königshütte (July),
      Friedenshütte (June).

1885 Hasper Eisen- und Stahlwerk, Haspe

1886 Hauts Fourneaux et Forges de Dudelange, Düdelingen

One of the last licensees was Röchling at Völklingen in January 1889. He did not previously need a licence of his own, since he had cooperated with Rheinische Stahlwerke,[39] where he was also a member of the board of directors.

When the German production of ingot steel increased from 233 049t in 1871 to 598 597t in 1880 (+157% or ll.1% per year) to 1 458 533t in 1890 (+144% or 9.3% per year, or nearly 10% per year over a period of 20 years!!),[40] this was due exclusively to the Thomas steel process, because in the same period the production of acid Bessemer steel decreased from 678 953t to 350 862t in 1890, and that of wrought iron remained more or less stable in the same period in absolute figures, (Fig. 12 and Table 3).

As an episode, one serious accident in the history of Thomas steelmaking should be reported: on the 20th of June 1894, at 9.15 p.m. the whole building of the converter steel melting shop in Hörde collapsed, killing three people. The reason was that the amount of dust accumulated on the roof became too heavy, perhaps together with rain which was absorbed by the thick layer of dust. The author believes that in the whole history of steelmaking a similar accident happened only once in later times, in the 1960s at Rourkela, India, which was called in the Indian press 'the famous roof collapse.' Nobody this time was killed, because nobody was present!

The author also remembers, that at the time of intensive Thomas steel production the so-called 'roof dust' was regularly collected and represented one of the materials included in the charge, thus improving the metal yield. In the beginning of the year 1882 the total number of Thomas converters was 25 in Germany, against 9 in Great Britain; one year later the number increased to 41.

One item, which was bitterly discussed, was that of quality and which products the Thomas steel could be used for. The main product of the Bessemer steel was rails, because that was the main product of the iron and steel plants, since railways at that time were exploding all over the world. Professor Tetmajer, from the Zürich Federal Polytechnic Institute (ETH) published a paper in 1894 'About the behaviour of Thomas steel rails in operation'.[41] He could be considered a neutral, since Switzerland had neither acid Bessemer steel works, nor

Fig. 12   Thomas steel production in Germany.

basic ones. He reported about the general situation of Thomas steel production, including those plants who never took up and those who gave up this process. So Krupp, who did not use the Thomas process at that time, (later only in 1898, a Thomas steel works was built in the newly founded Friedrich Alfred Hütte in Rheinhausen) bought minette deposits in Lorraine, and Witkowitz in Moravia gave it up, because the locally available iron ore was too low in phosphorus for the process and could only be used as long as the dumps of old P–rich puddle slag existed to increase the P level in the pig iron. He pointed out the importance of the recarburising procedure applied in several Thomas steel works (e.g. the process originally developed for the Bessemer process by Darby and another process developed by Spanagel at the Phoenix works, Laar and by Meyer, Düdelingen). He discussed all metallurgical aspects of the product and the experiences of the railway companies in Italy, Switzerland, Germany, Finland and Hungary. He came to the conclusion that basically the Thomas steel was completely fulfilling the requirements of the railways companies. Negative cases were due to negligence of the producer in executing the process and not due to a deficiency of the process itself. Much later it was learned that the Thomas steel even had an advantage in the case of rails: due to the higher nitrogen content, the resistance to abrasion was even better.

Thomas himself visited quite frequently his licensees, perhaps more often than Burnie mentions in his book, as can be seen in documents in the archives especially at Thyssen and Hoesch (Hörder Union) and apparently not always well planned ahead, like the following: 'Come with Thomas tomorrow'.[42] His visits at the works were always very friendly, like that described by Burnie[43] 'Got to Neuen Kirchen (Neunkirchen, Sarre) at 5 last night. Deputation to meet us at station; did works. Dined with owners, then beer and wine with all engineers till 12.'

THE THOMAS PROCESS IN THE AUSTRO-HUNGARIAN EMPIRE.

The introduction of the basic Bessemer process into Austria is quite interesting. Also in Bohemia there were preludes to the Thomas process. J. Jacobi succeeded in 1865 in leaching phosphorus containing ores by the application of sulphuric acid. After roasting he achieved

Table 3   Basic Bessemer steel production in Europe 1878–1900

	UK	German Empire	Austro-Hungarian Empire	France	Belgium	Sweden
1878	20	0	0	0	0	0
1879	1 150	1 782	3 500	0	0	0
1880	10 000	18 180	17 835	0	3 295	0
1881	46 120	200 000	31 889	12 306	14 200	0
1882	109 364	235 132	57 714	38 299	16 672	0
1883	122 330	328 909	88 429	113 000	27 399	0
1884	179 000	440 000	70 987	130 582	31 700	0
1885	145 707	548 212	76 821	122 711	21 056	0
1886	258 466	784 212	105 839	210 301	27 938	0
1887	435 046	1 167 702	118 379	222 333	50 777	0
1888	408 594	1 137 632	139 127	222 392	31 947	0
1889	493 919	1 305 887	126 502	240 638	47 037	0
1890	503 400	1 493 157	103 180	255 401	46 445	0
1891	436 261	1 770 779	95 061	287 528	38 793	0
1892	406 839	2 344 754	100 841	263 017	56 274	0
1893	385 036	2 342 161	108 104	*	*	8 419
1894	402 085	2 520 396	133 131	*	*	10 954
1895	448 782	3 004 615	127 816	*	*	17 824
1896	465 596	3 234 214	157 216	*	*	21 675
1897	517 973	3 606 734	*	*	*	26 373
1898	512 200	3 973 225	*	*	*	29 194
1899	525 657	3 973 225	*	*	*	28 933
1900	•	*	*	535 059	*	*

Source: Ludwig Beck: Geschichte des Eisens, Braunschweig 1903

a reduction of the P–content from about 2% to the range of 0.24 to 0.86 %. In the sense of the acid Bessemer process, he was not successful, not only because the elimination of phosphorus from the ore was not sufficient (only 70%) but also because the cost was extremely high, so that it was not suitable for use for Bessemer pig iron; but this process made some Bohemian ores (from Nucice) usable for the puddling process.[44] K.V. Zenger, professor of experimental physics at the polytechnic school of Prague, made some efforts in the direction of the later Thomas process but really did not succeed,[45] though he applied for a British patent in 1872 under the number 1063.

The patent for the Austro-Hungarian Empire had been applied for by Thomas via his patent agent Mr Paget in Vienna and was granted on the 20th of April 1879 under the condition that the patented process, according to the Austrian patent regulations, had to go into operation within the national territory before the 1st of April. This turned out to be a problem for Thomas. He finally asked the Adalbert Hütte (which belonged to the Prager Eisenindustrie with headquarters in Vienna) at Kladno in Bohemia by letter of 13th of January 1879 to make such melts within the foreseen time limit. He mentioned former trials in Belgium at Thy le Château in a Ponsard furnace and also mentioned trials in the USA. Somewhat later (16th of February) he required that pig iron with 1.5% Si and 1% P should be used with an addition of 120 kg lime together with iron ore. Prager Eisenindustrie confirmed in writing to Thomas on the 7th of January 1879, that they were ready to do so under the condition that they were granted free use of the patent rights and received 5% of the licence fees collected in Austria. This was confirmed by Thomas and reconfirmed by Prager Eisenindustrie on the 18th of February 1879. Kladno also confirmed on the 20th of April that the required melts had been produced for inspection by the government.[46] According to the Commission Report by Tunner, these had been made in the middle of March 1879.

There is a letter in the Hoesch Archives in Dortmund of Prager Eisenindustrie, dated 4th of June 1879, stating that Hörde apparently did not know about the agreement between Thomas and Prager Eisenindustrie.[47] Thomas answered to Prager Eisenindustrie on request that the rights of Kladno were safeguarded in the contracts with Massenez, who owned the patent rights for Austria (the respective contract could not be found). In this mess finally Thomas requested both parties to reach an agreement between themselves. Apparently he himself had difficulties in undoing the knot he had produced.

On the 5th of July 1879 it was agreed between Prager Eisengesellschaft and Hörde that Kladno had a free right to use and was entitled to receive 2.5% (instead of the former 5%) of the licence fees collected in the Austrian Empire and would be informed about all progress made in Hörde with the application of the Thomas process. They also exchanged experiences with bricks. Kupelwieser from Bohemian Ironworks asked for the results of the use of bricks produced by Vygen in Duisburg and confirmed that no new results were available there, because of new constructions in the works. The first trials at the Witkowice works were made in August 1879.[48, 49]

In July 1879 Kladno sent reports to Hörde on tests in a small Bessemer converter ('Swedish furnace') with pig iron of the following analysis: C 3%, Si 3.15%, P 0.53%, and another test with four heats in a 'big' converter of 5.2 tons and an analysis of: C 3%, Si 2.6% and P 1.3%. The P–content of the steel made was 0.23, 0.255, 0.030, 0.401 and 0.030%. The low phosphorus steels gave excellent values of tensile strength and reduction in area, the higher ones did not. The tests also show that the process later developed at Hörde, (that the Si–content should be lower and the P–content higher), was not yet known. Also, instead of dolomite, lime was used for the lining.

Already in July 1879 a most interesting book had been published by von Ehrenwerth, assistant professor at the mining academy of Leoben, on the technology and chemistry of the Thomas process.[50] This was an extended reprint of an article[51] published in April and June 1879 in Österreichische Zeitschrift für Berg- und Hüttenwesen. It is impressive how von Ehrenwerth analysed scientifically the Thomas process, even before the first real production melts had been produced, one could say even better than Thomas himself knew theoretically his process. The economic consequences of the invention were in fact detrimental for the old Austrian part (inner Austria) of the monarchy, since there were only phosphorus free ores (Erzberg in Styria) and no mineral coal, so pig iron cost more than in Britain, but advantageous for the northern part of Austria, (actual Czech Republic) because of the existence of mineral coal as well as phosphorus rich ores. Long before the knowledge of the later experimental results at Hörde, he showed by theoretical calculations that silicon could and should be replaced by phosphorus, thus saving lime addition to the charge and extending the life of the lining. He proved thus, that it was most advantageous to use the much cheaper white pig iron with low silicon content. The only mistake he made was to believe that phosphorus would be chemically combined with iron oxide. Gustav Hilgenstock discovered mineralogically[52] in 1883 that phosphorus becomes combined not with iron oxide but with lime, in the form of calcium triphosphate, which in the beginning von Ehrenwerth refuted, but the fact was chemically proved.

Another Austrian publication of great importance was the report composed by Professor Tunner in 1880.[53] The board of directors of the berg- und hüttenmännischer Verein für Steiermark und Kärnten decided to send a commission to visit the Thomas steelworks of continental Europe and study their results and experiences. Since this was considered of public interest, the travel expenses for one member of the group was paid for by the government. Six members started on the 13th of April 1880 from Leoben and came back after 3000 kms of travelling on the 29th of April. The commission visited the Thomas steelworks of Hörde, Ruhrort, Witkowitz, and Kladno, and also the Bessemer plants at Dortmund (Union), Phöenix in Ruhrort and Triniec in Austrian Silesia.

At Kladno the commission learned that they had built two cupola furnaces for the remelting of the pig iron and had ordered bricks at the newly built plant at Duisburg. They intended to go into operation with the basic converter by the end of May. The members decided that Kladno would be a very suitable place for the basic process (Fig. 13).

The bottoms at Witkowitz were made with tuyeres, whilst in most other plants the bottoms were rammed with iron needles (which by the way turned out to remain the most used procedure until the end of this steelmaking process in the 1960s). It was interesting to learn that they added phosphorus slag (from the converter process or the puddling process') to increase the P–content in the pig iron, because the local ores contained insufficient P.

At Hörde the commission was informed that not only German, but also English, French and Austrian metallurgists frequently visited the plant to study the progress achieved there. It was interesting to observe that the limestone addition to the charge was made in Hörde via a chute directly from the railway wagon instead of shovelling it in by hand. A still unsatisfactory situation was reported in the replacement of the bottoms. Both plants at Hörde and Ruhrort were trying to use the system invented by the American Holley to change bottoms completely and were hopeful that they would to be able to apply this system in the future. At

Fig. 13　Converter steelworks at Adalbert-Hütte in Kladno, Bohemia.

Hörde, Massenez normally boasted in the literature of the high quality of his basic steel, but at Ruhrort they used mainly inferior, cheaper pig iron for the Thomas process, thus preferring a cheaper product for simpler uses, and restricted the high quality raw material for better products.

The final conclusion of the commission's report was that the evaluation of the Thomas process for Austria was not as favourable as was the report by Tunner years before for the acid Bessemer process and less favourable than in Germany. The reason was very simple: the introduction had some advantages for northern Austria (later Czechoslovakia) with its reserves of mineral coal, but not for southern Austria (that is today's Austria) with its phosphorus free iron ores with low sulphur and no mineral coal. So Tunner's report confirmed the results of von Ehrenwerth's theoretical conclusions one year before. From the historical point of view this report is of high value for its description of different European works.

Nevertheless during the Vienna meeting of the Iron and Steel Institute in 1882 Thomas was awarded an honourable mention by the Prager Eisenindustrie-Gesellschaft.[54]

At about the same time an article was published in Pribram, Bohemia, on the question of the usefulness of the process for Bohemia.[55] The conclusion was not very clear. The author mainly saw advantages for the big plants with their own mineral coal resources, but not for smaller plants which were dependent on charcoal. Another advantage would be the lessening of the dependence on the import of foreign iron ores. Altogether this was a little shortsighted, as compared to the evaluation by von Ehrenwerth.

The famous Professor Tunner (letter to Massenez of 7th of November 1879[56]) not only in his commission report, was very doubtful about the functioning of the Thomas process, at least in the beginning. On October 28th of 1879, that is one month after the successful start of production in Hörde, Massenez wrote a long letter to his former teacher and friend explaining the perfect functioning of the process and inviting him for a visit.[57]

The details of the further successful expansion of the Thomas process in Bohemia, Moravia and Austrian Silesia cannot be described here. Documents on this subject can be found in the Hoesch Archives at Dortmund. Witkowitz in Moravia, Teplice in Bohemia and Triniec were further steps on this route. Carl Wittgenstein, initially the owner of Teplice, became later as head of the Prager Eisenindustrie one of the most influential personalities of the Austrian steel industry. His main opponent was Jan Dusanek from the Böhmische Montan-Gesellschaft.[58]

THE THOMAS PROCESS IN FRANCE AND BELGIUM

The French metallurgist Gruner published in 1859[59] an article explaining very correctly the metallurgy of phosphorus in steel; that due to the the easy reducibility of phosphorus pentoxide, phosphorus is more resistant than silicon to the oxidising effect of the iron oxide. Literally in the same journal in the following year (1860) on page 570, he said that a special, stronger reagent body is needed for the elimination of P from the iron. In 1865 Lencauchez proposed the puddling of P rich iron in a furnace lined with lime rammed with tar! In further publications Gruner dealt intensively with the physical-chemistry of P in iron, researching theoretically the process, but without finding a practical solution.[60] Gruner, professor at the Ecole Superieure des Mines, continued to investigate the theoretical details of dephosphorisation. He discussed other proposals like the elimination of the slag before the end of the decarburisation period (Prof. Wedding, Berlin) or the application of hydrogen or overheated steam.

On the 12th of March 1869 a French patent was granted to Emil Muller proposing a lining of magnesite for the steel furnace. Experiments came to an end in 1870, during the siege of Paris by German troops. Gruner again wrote in his Traité classique de Metallurgie in 1875 that possibly dolomite could respond to the metallurgical requirement of a basic slag. So it seems that French metallurgists had been extremely close to the solution of the problem. Interestingly enough they approached the problem from the scientific side, not from the craftsman's aspect.

In June 1879 the French metallurgist, Pourcel, of the Terre-Noire works read a paper about the Thomas process which attracted much attention, and was translated into German.[61]

Henri Schneider of Le Creusot had the first contacts with Thomas. He accepted the conditions for a licence and one converter was installed.[62] At Le Creusot difficulties arose, since they had in the beginning only 0.9% P in the iron, which was later increased. But the process was of limited interest for him, due to their ore situation. That was quite different in the north; Schneider informed de Wendel about the subject and Thomas concluded a contract with de Wendel in November 1879, but the plant at Joeuf was only inaugurated in 1883. After several negotiations, a Belgian ironmaster, Taskin, bought the rights for Belgium and the rest of France. De Wendel's rights were restricted to Meurthe-et-Moselle. These connections seem a little bit obscure and should be investigated in detail.[63]

In 1888 Thomas steel was produced in five plants:

Joeuf with 6 vessels of l0t each
Longwy with 3 vessels of 15t each,
Valenciennes with 2 vessels of l0t each,
Creusot with 2 vessels of 10t each
Pagny sur Meuse with 2 vessels of 10t each.

The expiry of the patents in France caused the construction of many plants, like:

Micheville and Pompey in 1895,
Frouard in 1900,
Homécourt in 1902
Neuves-Maisons in 1903,
Senelles in 1910,
Rehon in 1911,
Longwy Gouraincourt (La Chiers) in 1913.

The increase of Thomas steel production in France from 1884 to 1899 was 10.9% per year (+374% in 19 years), from 1899 to 1913 12.6 % (+325 % in 14 years). Thus the expansion of the Thomas process was somewhat slower in the first period in France than after the expiry of the patents in 1893.

The Company Cockerill in Seraing, near Liège in Belgium acquired a concession for the Thomas patents and made the first trials, but astonishingly it took more time in this country to come to a similar rate of increase to that in France and Germany, though the phosphorus rich ore deposits of Lorraine and Luxemburg were close to the steel plants. So the production of Thomas steel in Belgium in the first years was as follows:

1880   3295t
1881   14 200t
1882   16 672t
1883   27 399 t

The company of Rossius, Pastor & Co.[64] in Angleur was the first to take up real production in 1880, next was Souheur, Orban & Co. in Ougrée in 1881, and then Athus. From then on the production rose much faster: in 1906 the Thomas steel production was about 800 000t and in 1912 about 2 M t.

THE ENTRY OF THE THOMAS PROCESS INTO SWEDEN

The Bångbro works undertook the first trial melts in 1880[65] and bought the patent rights for Sweden in 1881. But the Thomas process did not come into commercial production until

1891 and then it was not at Bångbro's, who owned the patent, but at Domnarfvet, which belonged to the old Stora Kopparbergs Bergslags Actiebolag company, Falun. Domnarfvet built a new steel works with two acid Bessemer and three basic converters, which was commissioned by the German engineer Ferdinand Vahlkampf in October 1891.[66] Another Swedish steel company which tried to use the Thomas process was Stjernfors-Ställdalens verk in Örebro county, but only between 1896 and 1897. The patent owner Bångbro started the production of Thomas steel only in 1896. In much later years the so-called KalDo (Kalling–Domnarfvet) rotor vessel was developed at Domnarfvet by Bo Kalling. It was meant as a procedure to challenge the LD–process, but did not succeed.

So it is interesting to state that the Thomas process in Sweden, the country with the largest resources of phosphorus rich iron ores, with a much higher iron content than the minette (minette 30 to 35% Fe, Kiruna ore more than 60% Fe) did not exploit the Thomas process to the extent that other countries like Germany, France, Belgium and Luxemburg did. The reason is that Sweden practically had no mineral coal and was thus obliged to stay on special steels where the charcoal pig iron was advantageous and to refrain from mass steel production. So Domnarfvet produced 47 000t of Thomas steel in 1910, out of which 15 000t was on charcoal basis and 32 000t on the basis of coke pig iron. In 1913 it was 5 000t of Thomas steel with charcoal and 60 000t with coke. Since charcoal was much more expensive than coke, and coke had to be imported, this steel as a mass steel was certainly not competitive as compared to British and continental producers. The other side of the coin was the fact that Sweden exported for many years iron to Sheffield, which was produced in Lancashire fineries, since the charcoal iron was very much needed by the Sheffield cutlers for quality reasons.[67]

The main influence on the Swedish economy by the Thomas process was indeed not steelmaking, but ore supply. For many years up to 90% of the Swedish iron ore production was transformed into Thomas steel in Germany, thus having a major impact on the Swedish national economy.[68]

THE HOT METAL MIXER AS AN IMPORTANT CONSTITUENT FOR THE THOMAS STEEL PROCESS

In the early years the pig iron was remelted in cupolas, following the practice which had evolved for the acid Bessemer converter. The reason was that some plants had no blast furnaces at all and had to purchase pig iron from outside, or had not sufficient iron available in their own plant. Krupp, for example, bought a blast furnace plant at Sayn, some 150 kms away. Another reason was that the uniformity of composition from one tap of the blast furnace to the next was not satisfactory, though some plants used the direct conversion, so the cupola was needed for several reasons even though it added extra cost for fuel and handling. Until 1890 both procedures were in use for 'direct converting', that is pouring the liquid hot metal directly from the blast furnace into the converter and remelting in a cupola.

But there was another aspect of this problem: Gustav Hilgenstock, at first blast furnace manager at Hörde, later technical director, observed during the transport of the hot metal ladles towards the steel works often an acrid smell of sulphurous acid fumes, mainly with basic pig iron (manganese above 2%). The manganese rich iron showed at the surface, when cooling, a greenish slag which was composed of manganese sulphides. This kind of iron never showed problems with hot shortness (sulphur). Since basic pig iron generally contained less manganese, the steel quite frequently became red short. Thus it could be stated that a certain movement and delay of transport improved desulphurisation in the presence of a sufficient amount of manganese. With higher manganese contents the acrid smell appeared also with Thomas pig iron and green areas in the slag could be seen on the surface. So the reason for his patent application for the hot metal mixer was not only as a storage and mixing unit, but also as a desulphurisation apparatus, mixing a high manganese pig iron with the normal white iron for the Thomas process. The patent application was presented in May 1890. Hilgenstock applied for this patent without knowledge of an American invention of such a vessel for the mixing of several blast furnace taps, which was in use from 1884 at the Consett Works of Carnegie near Pittsburgh. A mixer had been in use since 1889 at the Edgar Thomson Works in Braddock also near Pittsburgh, . In the same year John Thomas King applied for a patent for such an apparatus in Liverpool. The English and American idea was mainly for the accumulation of hot metal, the German mainly for the desulphurisation.[69] Some examples of the desulphurisation effect of the mixer are:

blast furnace : 0.37 % S   0.058% S in the mixer
0.129%S   0.038% S
0.143%S   0.035% S
0.217% S   0.059% S

This was a reduction of the sulphur content by nearly 80%, which was a considerable result.

The Verein Deutscher Eisenhüttenleute (German Iron and Steel Institute) acquired the American patent from Carnegie, so the mixer could be used by all German steel companies without extra charge.

The design of the mixers in the beginning varied considerably, tilting vessels (like very big converters, Fig. 14) and flat pans, similar to an open-heart furnace were built. But finally the cylindrical, rotating form prevailed (Fig. 15). This kind of mixing vessel, heated with blast furnace top gas,[70] was rotated to fill with the hot metal from the blast furnace and to tap into the charging ladles of the steel works.

The size of the mixers also increased from originally ca. 80t, to 250t in 1898, to 500t in 1903, 900t in 1907 and finally 2 000t in 1914. A book of reference in 1961[71] recommends for a Thomas steel plant one to three mixers of 800 to 1500t capacity.

THOMAS SLAG, AN IMPORTANT FERTILISER.

In the beginning the slag produced by the Thomas process was mainly dumped, since the blast furnaces did not accept it due to the low content of iron oxide. In some works it was used in the blast furnaces to enhance the phosphorus content, where the natural content in the ore was too low for an appropriate Thomas process (chemically cold pig iron), for example in Bohemia. At Kladno a leaching treatment of phosphorus ore had been practised in

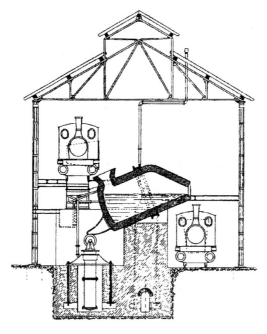

Fig. 14   Pig iron mixer at Hörde in the form of a converter.

Fig. 15   Pig iron mixer at Ruhrort.

1865 (see the section on Austria) to reduce the P content of the ore. The extracted phosphorus was sold from 1874 onwards for fertilising purposes with about 30 to 40% phosphoric acid under the trademark of 'Kladnophosphat'.[72] The product was equal in agricultural tests to superphosphate (a chemically treated natural mineral resource).

Thomas himself from the very beginning thought about the utilisation of the slag from his process. In Burnie's[73] book we read:

> 'The slag matter,' says his sister, 'tormented him. How right he was as to the capital importance of this question will be seen, when I state that, in 1889, 700 000 tons of basic (or 'Thomas') slag were produced (containing thirty-six percent of phosphate of lime), and that most of this immense quantity of slag was used as fertiliser, being applied directly to the land as a manure.
>
> 'In the winter of 1883–84, this valuable product was looked upon in England as so much mere troublesome rubbish, to be got rid of it somehow – by stacking, on waste ground – or even by taking it out to sea in barges and there depositing it. In Germany things were more advanced. The mode of utilising slag, which has eventually proved commercially successful, grinding it to a fine powder, had already been tried on the oolithic ores of Ikert, at Peine, by Herr Hoyermann. About 1880 that gentleman had applied the grinding treatment to the puddle slag produced at the Peine works. On the great success of the Thomas process in Germany, Herr Meyer, Chairman of the Peine works, pointed out to Hoyermann the greater richness in phosphorus of the 'Thomas slag'.

Such slag was, therefore, substituted for puddle slag with thoroughly satisfactory results. In the winter of 1882–83 what was known as 'Thomas phosphate powder' was first tried out on the land as a manure and in November 1883 Herren Hoyermann and Meyer were able to report to the German Royal Agricultural Society most excellent effects from its use, as Waggerman and Easterwood reported.[74] The first reference to Thomas slag as fertiliser found in the literature was in Germany in 1880[75] in a journal for the sugar beet industry.

How much importance was given in Germany to the slag is demonstrated clearly by the number of technical publications in *Stahl und Eisen*. In the period between 1881 and 1906 (25 years) 74 contributions had been published on the Thomas process itself and its technical problems, compared to 122 articles on the Thomas slag, its production elaboration and application.

In Table 4 it is amazing to see that the chemistry of the Thomas slags during 90 years of production has been quite uniform. Many procedures have been invented and patented to present the slag in a form which is most suitable for agricultural use, but most of them turned out to be too expensive.[76] The method most commonly applied was to dump the slag cakes, let them decay for some time, then crush them with a drop hammer and separate the iron inclusions by a magnetic separator (iron content ca. 4%). Finally the slag was ground extremely fine, by a ball mill or an edge mill, to make it as soluble as possible. Since the slag from Peine was very high in phosphate (the Peine hot metal contained ca. 3% P), the company of G. Hoyermann in Nienburg, Lower Saxony, was one of the first to market on a larger scale the Thomas slag as a fertiliser (see Burnie's book).

In the 1880s and 1890s many chemical and agricultural scientific authorities dealt with Thomas slag. In the beginning the slag was sold according to its total P_2O_5–content, but from the 1st of July of 1895[77] onwards it was decided unanimously amongst producers and consumers to sell it on the basis of the solubility in citric acid, which remained the basis until the end.

Table 4   Chemical analysis of Thomas slags in Germany

	Fleisher[1]	Osann[2]	Hütte 1910[3]	Hütte 1961[4]
P_2O_5	17.25%	17.14%	19.57%	16–19%
CaO	48.29%	49.80%	48.08%	47–50%

1. Fleischer: Entphosphorung des Eisens, Berlin, 1886, 7.
2. Osann: Lehrbuch der Eisenhüttenkunde, Leipzig, 1926, 217.
3. Hütte: Taschenbuch des Eisenhüttenmannes, Berlin, 1910, 570.
4. Hütte: Taschenbuch des Eisenhüttenmannes, Berlin, 1961, 606.

Always since then the slag has been an important contribution to the economy of the Thomas process. Thomas himself said:

'However laughable you may consider the notion, I am convinced that eventually, taking the cost of production into consideration, the steel will be the by-product, and phosphorus the main product'

This, perhaps, is somewhat exaggerated, but the author remembers the strong efforts made in Germany and France at the time of the dying Thomas process to continue to produce a Thomas slag with the BOF process and the LD–AC–process (LD avec chaux – LD with lime), or ARBED – Dudelange developed in Luxemburg, is proof of that. People believed that without the credit for the fertiliser slag, the process could not become economic.

In the 1920s, in the time of enforced economy after the war, the price paid to the German steelmakers for Thomas slag was reduced by a levy in favour of those French steelmakers who had sold their slag to the German agriculture before the war, since French agriculture had no market for Thomas slag.[78]

The slag mill also became an integral part of the Thomas steelworks. Because the sales area for fertiliser slag was a totally different sector (the language spoken by farmers and steel consumers was different) than the normal sales department of a steel company and for better control of prices, separate common organisations were founded. The first was founded in 1887. A successor was created in 1896, where the Prager Eisenindustrie (Kladno, Bohemia) took part. They separated from this Berlin based company in later years, and worked separately with a Bohemian Company. The sales organisation owned foreign agencies, because this had become an internationally traded product. In April 1939 a new company was founded[79] which included practically all German Thomas steel producers.

In the last phase of the Thomas steel production, the Thomas slag faced a marketing problem: for better solubility, the slag was ground as fine as flour. This caused some inconveniences during the distribution on the fields. So much research effort was devoted to investigating whether a granulating of the slag after grinding was possible without deteriorating the fertilising value of the product. But when the problem seemed to have been solved (in the sixties), the Thomas process ceased to operate.

DESIGN OF THOMAS STEEL PLANTS

The first Basic Bessemer plants were just remodelled Acid Bessemer Plants, so they were almost exactly the same as those works. The main difference in plant design was the 'dolo-mite plant' (Fig. 16) and the 'slag grinding and packing plant'. These sections were new to the plant arrangement and extra space had to be provided plus the transport facilities. In this sector a few companies specialised; for example the Laeis Company in Trier supplied all kinds of machinery needed in dolomite plants. A study prepared in 1956 shows the main steps of development.[80]

The original principle for both acid and basic Bessemer plants was the circular arrange-ment. One of these circular plants, erected and commissioned in 1884, continued in produc-tion until 1961 (nearly 80 years) with a total production of almost 18M Tons. This plant, at the then Aktiengesellschaft Phoenix in Laar, Duisburg, later Phoenix-Rheinrohr and then August Thyssen Hütte was the first to be built just for the new basic Bessemer process, according to the new requirements. The plant consisted in the first stage of three converters with a weight of charge of 10–13t. In 1906 a fourth vessel was added (Fig. 17). All cranes were hydraulic. In the middle of the converters a hydraulically lifted and turned working platform was mounted, which at the same time served as the crane for the pig iron charge to the converters and for the casting ladles for the tapping of the steel. This crane transferred the casting ladle to an adjacent similar casting crane, surrounded by the ingot moulds in a pit. This pit was similarly surrounded also by stationary hydraulic cranes for the manipulation of the ingot moulds and ingots. Before the introduction of industrial electricity, hydraulic power was the only practical form of energy in iron and steel plants. The first Bessemer plant of Krupp in Germany was of a similar design from 1862, and remained in service until the 1920s.

The next step, in the eighties of the last century, was to place the converters in a row. The pig iron was transported by steam driven cars upon an elevation, connected by a ramp with the shop floor, allowing the iron to be poured directly into the converters. The casting crane, designed as a steam driven car, below the converter stage, served the ingot molds arranged in a parallel row (Fig. 18). The general introduction of the pig iron mixer complicated the location of ramps for they had to travers from above the mixer to below this vessel with the result that they cut into unconnected parts of the area of the steel plant. Such a plant was also in service until the sixties with Phoenix-Rheinrohr at their Ruhrort plant in Duisburg (see Figs 7 and 8). None of these historic plants is still in existence, all of them have been disman-tled.

The next, and final step was the introduction of electricity as the main form of energy. Now the electrically driven overhead cranes became general (Figs 19–22).

It must not be forgotten, that an American mechanical engineer, A.L. Holley, who had already become famous by his contributions to the planning of acid Bessemer plants and the strongly discussed American successes in achieving extremely high productivity in Bessemer steelworks, outmatching by far European plants, made an important contribution to the plan-ning of Basic Bessemer steel plants, although the process was applied to a very limited extent in the USA.[81] He proposed to make the shells of the vessels interchangeable, leaving the trunnions fixed on the spot, so that the lining of the vessel could be replaced without

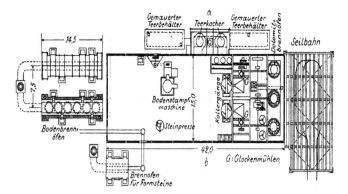

Fig. 16   Dolomite plant at Burbach (Luxemburg) 1908.

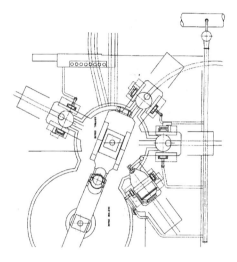

Fig. 17   Ruhrort plant of Rheinische Stahlwerke, built in 1884. In production until 1961. Total production 17.7 M t.

hurry in a separate bay. This procedure avoided the need for an additional converter to maintain the production of the plant. Two converters, were much easier to handle in production.

One case where this system was applied in Europe is reported by a letter of the managing director of the Witkowitz Steelworks in Moravia, Paul Kupelwieser, to his friend, the managing director of Hörder Bergwerks- und Hüttengesellschaft, Josef Massenez, that he intended to adopt Holley's system of interchangeable converters instead of building a 3rd and 4th one.[82] This plant was probably (as far as the author could find out), the only one built according to Holley's two converter plan (Fig. 23).[83] So it was a logical decision to apply also Holley's system of interchangeable converter vessels.

There is a certain parallel to the work of Taylor in the US around 1900, when he developed high speed steel and introduced the so-called 'Taylorism'.[84] His proposals were not much accepted in Europe.

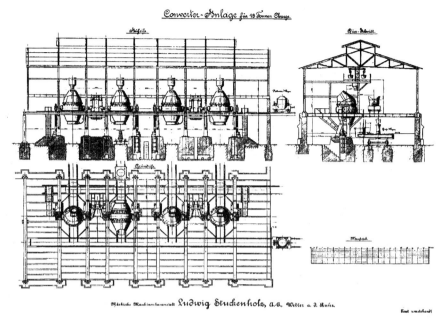

Fig. 18   Design of Thomas steelworks by Mr Trappen of Stuckenholz (predecessor of DEMAG), built for HADIR (Haut Fourneaux et Acieries de Differdange, later ARBED) in 1899.

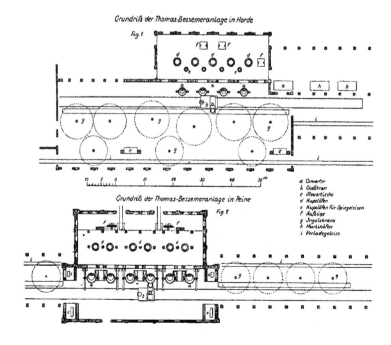

Fig. 19   Thomas plant of Höerde in 1882.

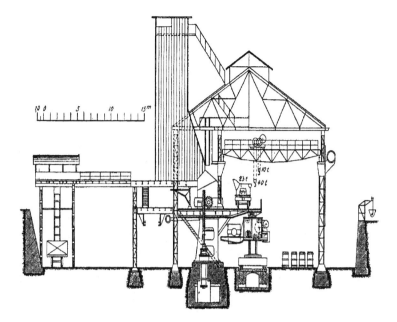

Fig. 20 Thomas steel plant of Rothe Erde (near Aachen), built in 1907.

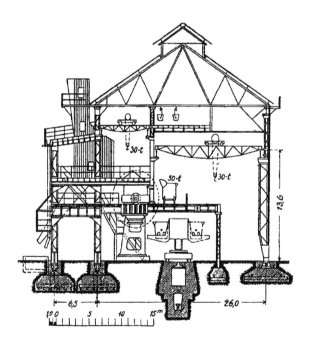

Fig. 21 Thomas steelworks of Esch (later ARBED).

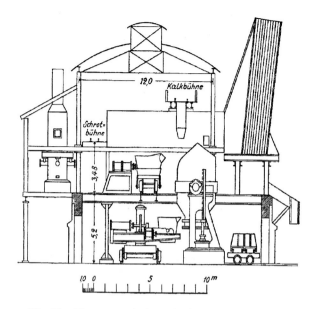

Fig. 22   Thomas steel plant of Friedenshütte in Silesia.

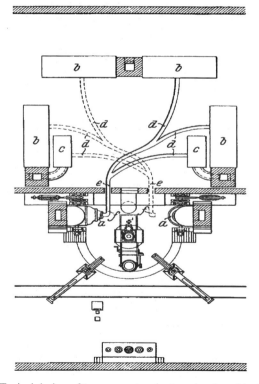

Fig. 23   Typical design of two converter plant, as developed in the USA.

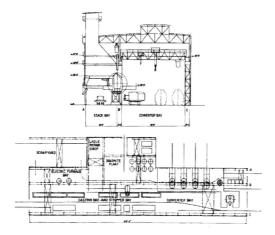

Fig. 24   Thomas steelplant built by DEMAG in 1956 in Helwan, Egypt.

An interesting contribution was made in Sweden in 1891[85] with a new device for the exchange of bottoms.

A very late Thomas steelworks was built at Helwan in Egypt by DEMAG about 1956 (Fig. 24).

The shape of the converter vessel was much discussed, but no scientific solution could be found. The shape of the burnt out vessel was used as a proposal, but this did not turn out to be an ideal shape. Practical experience showed only small improvements. The tilting equipment was hydraulic from the beginning (with the exception of the very first acid converters which were hand driven). This hydraulic movement remained till the end.

PROGRESS IN THE FINAL STAGE OF THE BASIC BESSEMER PROCESS.

Bessemer himself had proposed the use of pure oxygen in the pneumatic process. At that time this was totally uneconomic, since the production of oxygen was much too costly. This came nearer to reality only towards the end of the century by the inventions of Linde in Germany and Claude in France. Linde himself proposed the application of oxygen in the steel industry in 1905.[86] But even the refrigeration machine of Linde for the production of liquid air did not yet produce oxygen at an acceptable cost.

In 1908 Professor Wüst at Aix-la-Chapelle together with Laval[87] investigated theoretically the use of oxygen in the blast of Thomas converters, but still in the 1920s an application of pure oxygen in the steel industry was much to costly. A calculation, made in 1925,[88] showed a cost for oxygen of about RM 0.0575 per m^3. A price of below RM 0.02 was considered as the upper limit for the use in steel. It was discussed in the beginning mainly with regard to the application of oxygen in the blast furnace to increase the output. In the basic Bessemer process an improvement of cost was only thought about in terms of the possibility of melting an increased amount of scrap and obtaining a shorter blowing time.

Also possibly cheaper pig iron which would be chemically colder (e.g. a lower content of phosphorus) could be used. Metallurgical improvements were at that time not yet thought about.

In the year 1928 when the Linde–Fränkl process was introduced by Linde[89] the price of oxygen came down to RM 0.010–0015.

In 1925, tests with oxygen enriched air blast were made in the basic Bessemer plants in Oberhausen at Gutehoffnungshutte and in Dortmund at Hoesch.[90] Oxygen contents from 21% (air) up to 50% were tested. The blowing time could be reduced from 13.25 min under normal conditions to 5.56 min at 50% oxygen. The scrap addition to a pig iron charge of about 21 tons could be increased from ca. 600 kg to 3 500 kg. The metallic loss was independent from the oxygen content in the air. Interestingly enough, the question of steel quality was not even mentioned in these experiments. Only the economy by reduction of blowing time, thus increasing production and the savings by increased scrap addition, were considered. These factors were dependant on the price/cost ratio between scrap and pig iron and/or the possibility of refining pig iron with lower phosphorus content, which otherwise could not be used in the Thomas process. Another proposal for the improvement of the economy of the basic Bessemer process was the preheating of the air blast, which would facilitate the melting of an additional quantity of scrap in the converter at extremely low cost.[91] This proposal was just a theoretical one and was never realised.

In 1947, that is shortly after the war, at the Huckingen works of Mannesmann, work was done by W. Bading to improve the economy by oxidising vanadium from the pig iron into the slag. This was certainly important under the special postwar conditions.[92]

After the war especially in Germany, Belgium and Luxemburg,[93, 94] much effort was invested in improving the quality of the Thomas steel, since it was discovered that some of the disadvantages of the pneumatic steel as compared to open-hearth steel such as ageing, were due to the higher nitrogen content of this steel. Since in these countries the basic pneumatic process was still the major part of the production and the cheaper way of producing steel, several methods of reducing the nitrogen level in the steel were tested.[95] One procedure was the so-called HPN–process to decrease the nitrogen and phophorus content by adding oxygen to the air blast and adding ore to the melt (instead of cooling scrap), which replaced part of the oxygen from the air by oxygen chemically bound to the ore, so reducing the amount of nitrogen to be blown through the bath. Also the temperature was kept as low as possible. This procedure had been invented by Thyssen before the war. This and other procedures improved the quality of the Thomas steel, but were not yet sufficient to equal the quality of the open-hearth steel. Basic research was done by Oelsen and Geller.[96] These works defined the theoretical conditions to be observed to reduce the nitrogen level in the melt. Thus in the fifties several steel works in Belgium, Luxemburg and Germany added pure oxygen, steam or carbon dioxide to the air blast, and the addition of limestone (not calcined) instead of lime was investigated, which was introduced into the converter during the blowing process by use of a chute.

Important research and experimental work on the oxygen enrichment of the air blast was done also in Oberhausen with HOAG by Graef.[97] At the Oberhauesen plant of HOAG a Linde–Fränkl oxygen plant had existed since 1949, which was originally used for the blast furnaces, so oxygen was readily available. The results showed, that the economy of the application of oxygen in the Thomas process could be regarded as positive.

In Dortmund carbon dioxide was applied with much success[98] whilst steam was applied by Hoesch in Dortmund,[99] by ARBED in Luxemburg and Espérance-Longdoz in Seraing, Belgium.[100] At Espérance-Longdoz mixtures have been used which contained no air at all: 65% oxygen with 35% of steam. This mixture had about the same calorific effect as air, with the advantage of containing no nitrogen. With steam additions the nitrogen in the steel could be lowered to less than 0.0030%, and the P–content to less than 0.025%, thus approaching close to open hearth quality.

A probably last effort to keep the Thomas process alive was the so-called PL–process (Phoenix–Lanzen). In this process, during the last blowing time (the afterblow) no air was blown through the bottom, but pure oxygen was blown with a removable lance upon the surface of the bath. The advantage as compared to the Thomas and BOF process, was to be cheaper than BOF, and only a little more costly than Thomas, whilst obtaining a quality rivalling with the BOF steel.

In Germany the Thomas steel producing companies united in the so called 'Windfrischgemeinschaft' (Society of pneumatic steel producers) on the 1st of November 1951 to develop improvements of the process and interchange the use of experiences and patents etc., thus underlining the importance of the matter.[101]

All these efforts finally could not avoid the advent of the BOF pure oxygen process, which substituted completely for both Thomas and OH steel. The limit for all this most promising research was that the cost for open-hearth steel should not be exceeded.

CONCLUSION

The life cycle of the Thomas steel making process was exactly 90 years. Like in all other technical processes or equipments, towards the end of its life, the improvements in efficiency and quality were impressive and were never anticipated in earlier years . But this could not prevent the advent of oxygen steelmaking, which replaced all ingot steel processes invented in Britain in the 19th century; that is open-hearth, Bessemer and Thomas steelmaking. Strictly speaking the BOF is also an invention of Bessemer's but, due to the fact that oxygen gas at that time could not be produced at a reasonable cost, it could not be put into practice.

The Thomas process changed the steel industry on the European continent fundamentally: the Minette area, Luxemburg, Lorraine and the Sarre became important steelmaking regions only through the Thomas process, but also the Ruhr area owed a lot to Thomas, for although there was no phosphorus ore in the area itself, mineral coal was plentiful and water transport to the Ruhr area was easy. Sweden, though not one of the major Thomas steel producers, developed its northern mining area, Kiruna and Gellivara, which became the most important ore supplier to Germany. During some periods more than 60% of the German steel was made out of Swedish ore.[102] Certainly without the invention of Thomas the rise of Germany as a steel producer, for some time the third largest of the world, would not have been possible. The steelmakers of Germany, France and Belgium are greatly indebited to Sidney Gilchrist Thomas, who died in Paris on the 1st of February 1885, at the age of not yet 35 years (Fig. 25). His grave is on the cemetery of Passy and is still well kept by the city of Paris (Fig. 26).

Fig. 25   Sidney Gilchrist Thomas 1850–1885

Fig. 26   Grave of Sidney Gilchrist Thomas in Paris, cemetery of Passy.

In March 1885 the journal of the German Iron and Steel Institute published an obituary, from which a few sentences should be translated:

> *'On the 1st of February after extended sickness passed away in Paris our member Sidney Gilchrist Thomas at the youthful age of only 34 years. With sincere and deeply felt sympathy this news was learnt by the German metallurgists. He was much esteemed by many of them as a friend, appreciated for his personal friendliness, by all of them due to his extraordinary merits for the promotion of iron and steel metallurgy...*
>
> *In no other country the name of Thomas is as popular as in Germany, we always speak of the Thomas Process, and we have in our pig iron statistics a separate section with the title Thomas pig iron.*
>
> *His name is registered in the history of metallurgy in iron letters. He may rest in peace!'*[103]

ACKNOWLEDGEMENTS

The author is most grateful for the kind assistance extended to him by the directors of the Thyssen Archive, Dr Manfred Rasch, of the Krupp Archive Dr Renate Köhne-Lindenlaub, the library of VDEh, Dipl. Ing. Manfred Toncourt and their colleagues. All of them were most patient in attending the authors most exacting demands.

Last, but not least the author's gratitude is owed to Professor Bodsworth who undertook the painful and fatiguing task of transforming the text into readable English.

REFERENCES

1. O. Johannsen: *Geschichte des Eisens.,* Düsseldorf, 1953.
2. U. Troitzsch: Die Einführung des Bessemer Verfahrens in Preußen - ein Innovationsprozeß aus den 60er Jahren des 19.Jahrhunderts, Frank R. Pfetsch, Innovationsforschung als multidisziplinäre Aufgabe, Göttingen, 1975, 217.
3. M.J. Cournot: 'La fabrication de l'acier au convertisseur, Sa découverte - son introduc tion en France,' *Revue de Métallurgie* 20e année, 1923, N[0] 11 Novembre, 695.
4. C. Süterlin: *La Grande Forge*, La Coularde-Ile de Ré, 1981, 56.
5. B. Gille: 'Les problèmes techniques de la sidérurgie française au cours du XIXe siècle', *Revue d'Histoire de la Sidérurgie*, Nancy, 1961, Tome II, 45.
6. M.J. Cournot: 'La fabrication de l'acier au convertisseur, Sa découverte - son introduc tion en France,' *Revue de Métallurgie* 20e année, 1923, N[0] 11 Novembre, 695.
7. P. Tunner was ennobled because of his merits for the Bessemer Process in Austria.
8. P. Tunner *et al.*: Commisionsbericht über den Stand der Entphosphorung des Eisens im Bessemer Converter nach Thomas'-Gilchrist's patentiertem Verfahren, *Sonderabdruck aus der Zeitschrift des berg- und hüttenmännischen Vereines für Steiermark und Kärnten* 1880, 43.
9. B. Gille: 'Minerais algériens et Siderurgie metropolitaine', *Revue d'Histoire de la Siderurgie*, Nancy, 1960, Tome 1, 4.
10. J. Percy: *Metallurgy*, London 1864, **11** (3), 819.

11. Peine later became especially interesting for its extremely high phosphoric pig iron and the steel slag with very high contents of phosphoric acid, thus introducing the Thomas slag into the agriculture as a fertiliser.

12. Annual Report of the Secretary of the American Iron and Steel Association of 1889, 109.

13. G. Milkereit: 'Die Diskussion um das Moselkanalprojekt in seiner Verknüpfung mit der Einführung des Thomasstahlverfahrens in der westdeutschen Eisen- und Stahlindustrie 1883–1890,' *Tradition, Zeitschrift für Firmengechichte und Unternehmerbiographie*, l966, **1**, 184 and 232.

14. P. Tunner *et al.*: 'Commisionsbericht uber den Stand der Entphosphorung des Eisens im Bessemer Converter nach Thomas'-Gilchrist's patentiertem Verfahren,' *Sonderabdruck aus der Zeitschrift des berg- und hüttenmännischen Vereines für Steiermark und Kärnten,* 1880.
'30 Jahre Thomasverfahren in Deutschland', *Stahl und Eisen,* Jahrgang, 1909, **29**.
C. Balling: 'Welche Vortheile erwachsen Böhmen durch das Thomas-Gilchrist'sche Bessemerverfahren?,' *Oesterreichische Zeitschrift für Berg- und Hüttenwesen*, XXVIII Jahrgang, 1880, 45.
G. von Ehrenwerth: *Abhandlungen über den Thomas-Gilchrist'schen Process des Verbessemerns phophorhaltiger Roheisensorten*, Leoben 1879.
P. Trasenster: 'Déphosphoration des Fontes', *Revue universelle des Mines*, 1879, **6**, 225.

15. W.T. Jeans: *The Creators of the Age of Steel*, London 1884, 303.

16. R.W. Burnie: *Memoir and letters of Sidney Gilchrist Thomas, inventor*, London 1891, 124.

17. W.T. Jeans: *The Creators of the Age of Steel*, London 1884, 308.

18. C.-F. Baumann: *Zur Vorgeschichte des Deutschen Thomaspatents 12 700*, Duisburg, 1980.

19. J. Massenez:'30 Jahre Thomasverfahren in Deutschland', *Stahl und Eisen*, Jahrgang, 1909, **29**, 1466.

20. Later became part of ARBED.

21. Journal and Grand Livre de la Société Metz et Cie 1879–1880, Archives Industrielles Luxembourgeoises, Fonds MNetz - Eich - Dommeldange, ARBED, Un demisiècle d'histoire industrielle, 23, (unpublished commemorative work, Luxemburg 1964).

22. Personal opinion communicated by Dr Jacques Maas.

23. *Revue Universelle des Mines de la Metallurgie*, 1908, 165.

24. U. Troitzsch: 'Die Einführung des Bessemer Verfahrens in Preußen - ein Innovationsprozeß aus den 60er Jahren des 19 Jahrhunderts,' Frank R. Pfetsch: lnnovationsforschung als multidisziplinäre Aufgabe, Göttingen 1975, 217 and 224.

25. J. Massenez: '30 Jahre Thomasverfahren in Deutschland', *Stahl und Eisen,* Jahrgang, 1909, **29**, 1466.

26. L. Beck: *Geschichte des Eisens*, Band 5, Braunschweig 1903, 650.

27. U. Troitzsch: 'Die Einführung des Bessemer Verfahrens in Preußen - ein Innovationsprozeß aus den 60er Jahren des 19 Jahrhunderts,' Frank R. Pfetsch, lnnovationsforschung als multidisziplinäre Aufgabe, Göttingen. 1975, 239.

28. *Neue Deutsche Biogaphie Bd.2*, Berlin, 1955, 39, and *Neue Deutsche Biogaphie Bd.* 18, Berlin, 1997, 736.

29. Contract between Rheinische Stahlwerke and Hörder Bergwerks- und Hüttenverein with three refractory manufacturers, Hoesch Archives, Dortmund.

30. L. Beck: *Geschichte des Eisens*, Band 5, Braunschweig 1903, 645.

31. '30 Jahre Thomasverfahren in Deutschland', *Stahl und Eisen*, Jahrgang 1909, **29**, 1465.

32. Akademischer Verein Hütte: Hütte, Taschenbuch für Eisenhüttenleute, Berlin 1961, 611.

33. Technischer Verein für Eisenhüttenwesen, Bericht über die General-Versammlung vom 14 December 1979 zu Düsseldorf.

34. see his paper read at the Iron and Steel Institute 1880: R. Pink: 'On the Dephosphorisation of Iron in the Bessemer Converter,' *Journal of the Iron and Steel Institute,* 1880, 57.

35. Newspaper message, Thyssen Archive.

36. Thyssen Archive 640 / 00

37. Peine had a special situation, the pig iron of the local ore resulted in an extremely high P-content, that is 3%! Peine obtained later a special importance for its slag.

38. L. Beck: *Geschichte des Eisens*, Band 5, Braunschweig 1903, 662 and Hoesch archives DHHU 775.

39. U. Wengenroth: *Unternehmensstrategien und technischer Fortschritt, Die deutsche und britische Stahlindustrie 1865–1895*, Zürich 1986,181.

40. H. Marchand: *Sakularstatistik der deutschen Eisenindustrie*, Essen, 1939.

41. L. Tetmajer: *Ueber das Verhalten der Thomassathlschienen im Betriebe*, Zürich, 1894.

42. Telegram from Massenez at Hörde to Göcke, chairman of Rheinische Stahlwerke at Ruhrort, 7th of April 1880, Thyssen Archives 640 000.

43. R.W. Burnie: *Memoir and Letters of Sidney Gilchrist Thomas, inventor*, London 1891, 161.

44. L. Jenicek and I. Krulis: *British inventions of the industrial revolution in the iron and steel industry on Czechoslovak territory*, National Technical Museum, Praha, 1968, and Archives of Kladno Iron works (Journal S/12).

45. L. Jenicek and I. Krulis: *British inventions of the industrial revolution in the iron and steel industry on Czechoslovak territory*, National Technical Museum, Praha 1968, and the correspondence of Zenger in the archives of the museum.

46. Hoesch Archives, Dortmund.

47. It seems that no formal contract on this subject exists, but only the above mentioned exchange of letters.

48. F. Kupelwieser: Resultate der in Witkowitz ausgeführten Entphosphorungsversuche des Eisens nach dem von Thomas und Gilchrist patentirten Verfahren. Österreichische Zeitschrift für Berg- und Hüttenwesen. No. 38, 1879, XXVII Jahrgang, 451, and P. Tunner *et al.*: Commisionsbericht über den derzeitigen Stand der Entphosphorung des Eisens im Bessemer Converter nach Thomas Gilchrist's patentiertem Verfahren, Sonderabdruck aus der Zeitschrift des berg- und hüttenmännischen Vereines für Steiermark und Kärnten 1880.

49. I. Krulis-Randa: 'The introduction of British metallurgical techniques into Austria', *J. of the Iron and Steel Institute*, January, 1960, 43.

50. G. von Ehrenwerth: Abhandlungen über den Thomas Gilchrist'schen Process des Verbessemerns phophorhaltiger Roheisensorten, Leoben 1879.

51. Österreichische Z. für Berg- und Hüttenwesen, Jahrgang XXVII, 1879, 277.

52. G. Hilgenstock: *Stahl und Eisen*, 1883, 498 and 'Über die Zusammensetzung der Thomasschlacke und deren Begründung', *Stahl und Eisen* 1886, 525.

53. P. Tunner *et al.*: Commisionsbericht über den derzeitigen Stand der Entphosphorung des Eisens im Bessemer Converter nach Thomas Gilchrist's patentiertem Verfahren. Sonderabdruck aus der Zeitschrift des berg- und hüttenmännischen Vereines für Steiermark und Kärnten 1880.

54. *Stahl und Eisen*, March 1885, **3**, 177.

55. C. Balling: Welche Vortheile erwachsen Böhmen durch das Thomas-Gilchrist'sche Bessemerverfahren? Österreichische Zeitschrift für Berg- und Hüttenwesen, XXVIII Jahrgang 1880, 45.

56. J. Massenez: '30 Jahre Thomasverfahren in Deutschland', *Stahl und Eisen*, Jarhg. 1909, **29**, 1466.

57. J. Massenez: '30 Jahre Thomasverfahren in Deutschland', *Stahl und Eisen*, Jarhg. 1909, **29**, 1466.

58. L. Jenicek and I. Krulis: *British inventions of the industrial revolution in the iron and steel industry on Czechoslovak territory*, National Technical Museum, Praha 1968.

59. Gruner: *Annales des Mines,* 1859, tome XV, 291.

60. M.L. Guillet: 'l'Historique des procédés basiques de fabrication de l'acier et Sidney Gilchrist Thomas', *Revue de Metallurgie XIV*, 1917, 3.

61. A. Pourcel: Réunion de la Societé de l'Industrie Minérale, le 7 juin 1879, translated into German for the members of the Technischer Verein für Eisenhüttenwesen, separately printed.

62. L. Beck: *Geschichte des Eisens*, Band 5, Braunschweig 1903, 1089.

63. J.M. Moine: L'adoption du procédé Thomas par la siderurgie lorraine. Sources nouvelles et conclusions définitives, 1996 unpublished.

64. The General Manager of the Ruhrort plant of Rheinische Stahlwerke (Acieries du Rhin, with headquarters in Paris, with French and Belgian capital was also called Pastor, and was one of the major shareholders.
C.-F. Baumann: Von der Stahlhütte zum Verarbeitungskonzern, Thyssen Industrie 1870–1995, Essen, 1995.

65. A. Attman: Svenskt Järn og Stäl. 1800–1914, Jernkontorets Brghistzoriska Skriftserie 21. Stockholm 1986.

66. 'Einführung des Thomas-Processes in Schweden', *Stahl und Eisen*, January 1892, 8.

67. F. Toussaint: 'Korså bruk, a Swedish Lancashire finery working until 1930', *Steel Times*, 1997, **225**, 424.

68. F. Toussaint: 'Sweden and Germany, a historical partnership in steel', paper read at the 250th anniversary of Jernkontoret in Stockholm, *Iron and Steel – Today, Yesterday and Tomorrow*, Vol.3 Stockholm 1997.

69. J. Massenez: '30 Jahre Thomasverfahren in Deutschland', *Stahl und Eisen*, Jahrgang, 1909, **29**, 1466; same article 1479, and Hilgenstock: *Stahl und Eisen*, 1891, 798.

70. B. Osann: *Lehrbuch der Eisenhüttenkunde Band 2*, 69.

71. Akademischer Verein Hütte e.V. Hütte, Taschenbuch für Eisenhüttenleute, Berlin 1961.

72. M. Fleischer: *Die Entphosphorung des Eisens durch den Thomas-Prozeß und ihre Bedeutung für die Landwirtschaft*, Berlin, 1886.

73. R.W. Burnie: *Memoir and letters of Sidney Gilchrist Thomas, inventor*, London 1891.

74. W.H. Waggerman and H.W. Easterwood (Washington DC): Basic slag as Phosphate Fertilizer, *Chemical and Metallurgical Engineering*, November 1923, 873.

75. A. von Wachtel: 'Eine Quelle für Phosphorsäure', *Organ Centralvereins für Zuckerrübenindustrie 18*,1880, 4.

76. M. Fleischer: *Die Entphosphorung des Eisens durch den Thomas-Prozeß und ihre Bedeutung für die Landwirtschaft*, Berlin, 1886.

77. 'Verkauf der Thomasschlacke nach Citrattlöslichkeit', *Stahl und Eisen*,1895, 5l9.

78. 'Die schädigende Wirkung der Höchstpreisermäßigung für Thomasphosphatmehl auf die Wettbewerbsfähigkeit der deutschen Industrie', *Stahl und Eisen*, 1921, **41**, 838.

79. Thyssen Archives, VSt. 1365.

80. F. Toussaint: *Studie zur Planung von Konverterstahlwerken*, Berlin 1956.

81. A.L. Holley: 'Adaptation of Bessemer plant to the Basic Process', *Transactions of the American Society of Mechanical Engineers*, Vol. 1, 1880, New York.

82. Letter from P. Kuppelwieser to Josef Massenez, 9/5/1882, Hosed Archives DOUGH.

83. Paul Kupelwieser: 'Der basische Besssemer-Process', (Paper read at the autumn meeting of the Iron and Steel Institute 1881 in London), *Stahl und Eisen*, Düsseldorf, November 1881, 181.

84. F. Toussaint: *High speed steel, one of the most important contributions of the steel industry to machine building in the last 100 years*, IDIOTIC Liege, 1997.

85. 'Einführung des Thomas-Processes in Schweden,' *Stahl und Eisen*, Nr. Januar 1892, 8.

86. E. Karwat: 'Fortschrittsbericht über die Anwendung von Sauerstoff für die Eisen- und Stahlerzeugung,' *unpublished memorandum*, Höllriegelskreuth/München, 1947.

87. *Metallurgie* 1908, **5**, 431/62, 471/89 and *Stahl und Eisen*, 1909, **29**, 121/33.

88. A. Brüninghaus: 'Die Gewinnung und Verwendung von sauerstoffangereicherter Luft im Hüttenbetriebe, *Stahl und Eisen*, Jahrg. 1925, **45**, 737.

89. E. Karwat: Fortschrittsbericht über die Anwendung von Sauerstoff für die Eisen- und Stahl erzeugung, *unpublished memorandum*, Höllriegelskreuth, 1947.

90. J. Haag: 'Die Verwendung von sauerstoffangereichertem Gebläsewind beim Thomasverfahren,' *Stahl und Eisen,* Düsseldorf, 1925, **46**, 1873.

91. Contribution to a discussion by A. Schack: *Stahl und Eisen*, 1939, **59**, 81 5.

92. W. Bading: 'Die Entwiscklung des basischen Windfrischverfahrens,' *Stahl und Eisen*, Jahrgang, 1947, **66/67**, 137.

93. P. Coheur: 'Le dévelopment de la recherche métallurgique en Belgique,' *Revue Universelle des mines 9ᵉ série*, T.VI, 1950, **8**, 203.

94. T. Kootz: 'Zur belgischen metallurgischen Forschung der Gegenwart,' *Stahl und Eisen*, 1951, **71**, 58l.

95. F. Toussaint: *Entwicklung und gegenwärtiger Strand des Windfrischens mit Gemischen aus Sauerstoff, Wasserdampf und Kohlendioxyd*, Berlin, 1956.

96. W. Oelsen:Physikalisch-chemische Grundlagen der Verfahren der Eisen- und Stahlerzeugung, *Stahl und Eisen*, 1948, **68**, 175 and W. Geller: 'Zur Theorie der Entgasung flüssiger Metallbäder durch Spülgas, *Zeitschr. für Metallkunde*, 1943, 213.

97. Rudolf Graef: 'Metallurgie und Wirschaftlichkeit der Sauerstoffanreicherung des Gebläsewindes beim Windfrischen,' *Stahl und Eisen*, Jahrgang 1951, **71**, 1189.

98. Meyer, Knüppel, Darmann and Pottgießer: *Stahl und Eisen*, 1952, **72**, 225 and 1409.

99. V.d. Esche: 'Das Frischen von Thomasstahl mit überhitztem Wasserdampf oder Sauerstoff', *Stahl und Eisen*, 1950, **67**, 322.

100. Coheur, Marbasi and Daubersy: 'Fabrication d'aciers Thomas de haute qualite', *Revue Universelle des mines*, 1950, 104, and Coheur, Marbasi and Daubersy: 'Fabrication et proprietés des aciers Thomas à bas azote', *Revue Universelle des mines*,1950, 401.

101. Archiv des VDEh, Düsseldorf and Thyssen Archiv Duisburg.

102. F. Toussaint: 'Sweden and Germany, a Partnership in Iron and Steel', *Iron and Steel - Today, Yesterday and Tomorrow*, Conference to celebrate the 250th anniversary of Jernkontoret. Stockholm, June 1997, To be published.

103. *Stahl und Eisen*, March 1885, **3**, 177.

A Bessemer Miscellany

T. J. LODGE

THE BESSEMER STEEL PATENTS LICENSORS

It is not widely appreciated just how much Henry Bessemer's early ventures into the steel business were family matters. This applied in the case of the patent licensors, the Greenwich Works, and the company established in Sheffield to produce Bessemer steel on a commercial basis.

Henry Bessemer and his brother-in-law (and business partner) Robert Longsdon were the 'Registered owners and Proprietors' of the several patents taken out between 1856 and 1863 in connection with steel making and forming. The royalty paid by licensees varied between £1 per ton and £2 per ton of steel made, depending on the end use. Bessemer, in his *Autobiography,* mentions that at some stage he sold a 1/4-share of the rights to the patent royalties to John Platt, MP for Oldham, and a consortium of Manchester business men, but I have been unable to establish how long this new arrangement lasted.

Bessemer's main patent expired in 1869, and most of the other early ones had run out by about 1872. Possibly in anticipation of this, Bessemer registered a whole new string of patents in the late 1860s, which again were administered by 'Bessemer and Longsdon'. Examination of surviving copies of licences taken out in the period from 1870 onwards indicate that the patent royalties were still very much 'in the family'. Both Bessemer and Robert Longsdon signed the 1870 re-licensing for Samuel Fox, and interestingly, Robert's signature was witnessed by a David Longsdon. Robert Longsdon, who was an architectural engineer by profession, and a founder member of the Iron and Steel Institute, passed away in the early 1870s. Later licensings were signed by Henry Bessemer, his sister Maria (Robert's widow), Alfred Longsdon (Robert's brother) and David Longsdon. The indications are that Alfred was possibly acting in the capacity of a trustee of Robert's estate to safeguard the rights of Maria; David could well be the son of Robert and Maria.

We know more of Alfred Longsdon than Robert, and it is clear he played a key role in promoting the Bessemer process. He too had been a founder member of the Iron and Steel Institute, and as Krupp's English agent had been responsible for introducing John Ramsbottom of the London and North Western Railway in the late 1850s to crucible-cast steel locomotive axles and tyres produced by Krupp. He also played a part in persuading Ramsbottom to give Bessemer steel a trial a little later, resulting in the establishment of the Bessemer steelworks at Crewe. His links with Krupp help us to appreciate why Krupp was chosen by Bessemer to introduce the process into Germany.

Alfred lived at Denmark Hill in London, possibly even in the same house as Henry Bessemer as part of the 'extended family' that was so much a feature of Victorian England. Strangely, his obituary in the *Journal of the Iron and Steel Institute* (1894), while it tells of his connection with Krupp, gives no mention of Bessemer whatsoever!

THE GREENWICH BESSEMER WORKS

Another Bessemer family venture which has been virtually ignored by all modern writers is the Bessemer Works at Greenwich, London. This was established by Henry Bessemer on 3 acres of land on the south bank of the River Thames at Greenwich in the early 1860s with a view to putting his two sons into the steel business. At the time the Thames was a busy shipbuilding centre, and it was thought that this would provide the works with regular orders.

The plant, off Blackwall Lane in the former Greenwich Marshes area, was intended to operate with two 2 1/2 ton capacity converters and steam hammers for ingot cogging. It was styled 'Bessemer Brothers' in Kelly's 1869 *Post Office Directory for London*, and Morris' *Business Directory of London* of the previous year gives the works title as 'The London Iron and Steel Works' at East Greenwich. A contemporary list of UK Bessemer plants indicates it actually had two 5 ton converters. Richard Price-Williams was appointed Manager but for some reason Bessemer chose not to open the works for commercial production, so the project was still-born. Apparently Bessemer was convinced that the days of shipbuilding on the Thames were numbered, and his efforts should be concentrated in the UK's established iron and steel districts. His fears were a little premature, however, for the Thames Iron Works and Shipbuilding Co Ltd. on the Isle of Dogs (Millwall) survived into the present century, not launching its last Dreadnought until 1911.

After abandoning the Greenwich plant Bessemer let it to the Steel and Ordnance Company, which is shown in the *Journal of the Iron and Steel Institute* for 1871 as the Bessemer Steel and Ordnance Company, with two 4 ton converters. Doubtless these were the same vessels as in the earlier listing. In any event, this company – despite the proximity of its plant to Woolwich Arsenal – had little or no success with the venture, and later the steam crane builders Appleby Brothers occupied the Greenwich site for some years.

Today you will search in vain for Blackwall Lane, for much of its route is occupied by Tunnel Avenue, the southern approach to the Blackwall Tunnel, and the whole area is currently being redeveloped anyway with the building of the Millennium Dome. How appropriate it would have been, under the circumstances, to incorporate some Bessemer steel somewhere in the dome. Failing this, there really ought to be a materials display as part of the exhibits, with Bessemer steel prominently included. It would be a nice touch in memory of what almost became London's first commercial steel works.

MEMORIES

JOHN J BLECKLY

John James Bleckly was the last man alive who witnessed Bessemer's first public converter experiment at Baxter House in 1856. He recalled the occasion when writing to Joseph Butler Junior of Youngstown, Ohio, to thank him for sending a copy of *History of the American Iron Trade.*

In his letter, dated 9th November 1922, Bleckly, who was Deputy Chairman of the Pearson & Knowles Coal & Iron Co Ltd, wrote: 'One of my earliest recollections of the trade relates to the historical meeting of Ironmasters in 1856, at which I was present as a boy. The scene was a small iron foundry in the Edgeware Road in London, where Henry Bessemer – who was a friend of my father – gave the first public exhibition of the Bessemer process in the practical production before our eyes of a Bessemer steel ingot six inches long by four inches square – made in and cast from a small foundry Cupola, into which a special high-pressure air-blast was introduced when the metal was melted. The pig-iron used was Pontypool cold-blast, the best quality of iron known in those days. There were about 40 Ironmasters present, who represented all the leading works in the kingdom, and I am now the only survivor of that party. The process was condemned by the majority of those who saw the experiment as a failure, the material that Bessemer had to use to obtain his results – ie, cold blast pig iron – being far too expensive for ordinary purposes.

'Mr Menelaus, the manager of the famous Dowlais Iron Works and reputed to be the most skilful iron-maker of his day, thought otherwise, and agreed to put down the necessary plant to make a thorough practical trial of the new process on a large scale, on the condition that if he made 20 000 tons of Bessemer Steel successfully his Company should have the free use of the patents without royalty to the end of their term. A large sum of money was spent upon an exhaustive investigation of the capabilities of the process which, however, resulted in failure, and the free royalty arrangement finally lapsed. This happened, I may say, before Mushet ultimately succeeded in solving the problem by the introduction of manganese during the operation. I believe that there is now no one left in the world who witnessed this original trial or who can tell this tale as I can from personal experience'.

He continued by recalling an early episode in the history of the Iron and Steel Institute which involved both Bessemer and Siemens. 'I may here mention a curiously interesting incident which I remember, and of which I think I am also now the only person alive who witnessed it. It occurred in the very earliest days of the Iron and Steel Institute, I think in 1867. I was present at the Landore Steel Works in South Wales, which were erected by Siemens and some of his friends to demonstrate the advantages of the Siemens process on a commercial scale and were the first works of their kind.

'A test was arranged to be made in the presence of the Members of the Iron and Steel Institute, of which I was one, to prove the exceptionally excellent quality of Siemens steel rails, and a sample rail was broken by way of experiment. I was talking with Mr Henry Bessemer as we were watching the test, Mr Siemens being a short distance away, when the falling weight of a ton struck the rail to be tested; it broke off a piece weighing about half a hundredweight, which flew past us within a yard of Mr Bessemer's head. Had it struck him, he would certainly have been killed on the spot, and there would have been the tragedy to record - that the first public test of Siemens steel had killed the inventor of the Bessemer steel process - which Siemens was so anxious to supersede. Fortunately the broken rail missed the mark by a few feet, and no harm was done.'

'BESSIE' DAVIES

Through the recollections of others older than ourselves we are able to reach back in time and often discover facts which have so far escaped written records. Such is the fascination of oral history, a most useful means of research.

In June 1986 I was privileged to converse with Mrs 'Bessie' Davies, then a frail 101 years old lady living in a Sheffield old people's home. Mrs Davies, as a child, lived in Denmark Hill, London, in the house neighbouring that of Sir Henry Bessemer. (Her parents were in service at the time.)

She recalled that she often saw Bessemer returning home by carriage if she happened to be playing at the bottom of the drive, and he occasionally spoke to her, sometimes passing her sweets. Her major impression was that Bessemer, by this stage, was a rather sad old man, seemingly almost always preoccupied with his thoughts.

Bessemer died shortly after these encounters, and Mrs Davies ultimately came to Sheffield, the scene of Bessemer's greatest commercial success, by pure coincidence.

Mrs Davies was, in all probability, the last living soul who could personally recall Henry Bessemer, the man who gave the world its first viable process for making tonnage steel. Not an earth-shattering story, but a fact worth recording.

IN MEMORIAM

America did Bessemer proud. There are no fewer than thirteen places which feature Bessemer in their name, many being towns once associated with the Bessemer steel industry. As a result, there was once even an American railroad company which included his name in its title – the Bessemer and Lake Erie.

Great Britain's efforts, by comparison, were rather feeble. Several towns had – and a few still have – streets named after Bessemer. These were no grand boulevards named out of civic pride; rather mere residential streets built in the proximity of the local Bessemer Steelworks. Examples could be found in Liverpool, Manchester, Sheffield, Spennymoor and Stocksbridge. London does, however, have a Bessemer Road off Denmark Hill – particularly appropriate – and Rotherham has recently remembered its heritage by creating a Bessemer Way in the 'Bessemer Park' Industrial Estate built on part of the site of the original Ickles Bessemer Works. Cardiff and Newport, too, have used his name in recent estate developments which recall several great scientists/engineers.

One major UK railway company, the London and North Western Railway, went out of its way to remember its debt to Bessemer. Over the years, no fewer than three LNWR express locomotives built at Crewe carried his name – a 2–4–0 'Precursor' type built in 1879 ; a

2–2–2–2 'Greater Britain' compound built in 1894 ; and a 4–6–0 'Experiment' built in 1907. The transfer of the name as each type was superseded by a bigger and more powerful design was a rather nice touch. In view of Bessemer's overwhelming interest in developing new technology to supersede out-dated methods, one feels he would have approved of the LNWR's approach!

Several steelworks also remembered their origins by honouring Bessemer's name: Blaenavon, Bolton Iron & Steel, Glengarnock, Ebbw Vale and Dorman Long (as successor to Bolckow Vaughan and North-Eastern Steel) each named one of their shunting locomotives BESSEMER. Whoever at Dorman Long sanctioned the naming was well versed in the politics of compromise – another contemporary locomotive was named THOMAS, and SIEMENS followed shortly after! Finally, Wilson, Cammell & Co. Ltd. of Dronfield had a newly built locomotive of 1879 named SIR HENRY BESSEMER in recognition of his recent civic honour.

Oddly, the most important of the industrial locomotives named BESSEMER actually belonged to an iron producer which never became involved in Bessemer steel manufacture! As part of a commercial understanding between the London and North Western Railway and the Wigan Coal and Iron Company, the latter was allowed to adopt certain standard Crewe locomotive design features for a number of six-coupled inside cylinder industrial saddle tank shunting locomotives it built at Kirkless Ironworks at Wigan. In return, Wigan Coal and Iron allowed Crewe to use the first of these, assembled in 1865, as a proving bed for Bessemer steel. The frames, boiler and other important parts of the locomotive – fittingly named BESSEMER – were in all likelihood supplied from Crewe's own Bessemer steelworks. The locomotive, the first built essentially wholly of Bessemer steel, survived until about 1919, when it was scrapped. What a pity it wasn't considered a candidate for preservation!

Our acedemic institutions have fared little better than our civic fathers in remembering the great man. True, the Institute of Materials continues to award the Bessemer Gold Medal annually, but this invariably goes to a senior member of the profession. Apart from this we have no Bessemer College (compare Brunel University); no Bessemer 'chair' (professorship) within academia; and no 'polytechnic' Bessemer Prize for the most promising student, be he or she Civil, Electrical, Materials or Mechanical Engineer. Surely we can do better than this?

VARIATIONS ON A THEME

Introduction

No modern book celebrating Bessemer's metallurgical achievements would be complete without a consideration of the various second and third generation processes spawned by the basic concept of Bessemer's converter. By this is implied the use of a refractory-lined pear-shaped vessel mounted on trunnions, and capable of being tilted to facilitate charging, emptying, mixing contents and sampling.

The earliest and latest variations, coincidentally, both originated in America. The earliest, Kelly's Converter, appears to be outright plagiarism and in defence to Bessemer is not considered further here; the most recent, the AOD Converter, is responsible each year for producing million of tonnes of stainless steel – very much a material of the modern world.

PASTURES NEW

Possibly the first variation of the Bessemer process of real commercial significance was for the production of copper. Sulphide ores were traditionally smelted in a reverberatory furnace to concentrate the copper into matte and eliminate the gangue as slag. Prior to 1880 virtually all the world's copper matte was shipped to the Swansea region of South Wales for conversion to blister copper by a labour intensive process. Attempts were made in America to oxidise the matte in conventional Bessemer Converters but these failed due to the tuyères blocking with chilled copper. The difficulty was finally overcome by Pierre Manhes of France, who moved the tuyères to the circumference of the vessel (i.e. side wall), thus allowing the quiescent (refined) copper to settle beneath the blast. Even so, it was necessary periodically during the second stage of the blow to keep the tuyères clear by punching a steel bar through them. The first copper converter of this type was installed at the Parrot Works, Butte, in 1884.

STEEL FOR THE FOUNDRY

By the turn of the century, there was a whole family of Bessemer Converter variants dedicated essentially to the production of small quantities of soft low carbon steel almost exclusively intended for the manufacture of castings. They fall into two categories, fixed and rotating. The former were all fitted with tuyères in the vessel circumference (i.e. side wall) and provided with a tap hole at the base for tapping the refined charge into a ladle. The small rotating vessels could be equipped with tuyères either in the base (as conventional Bessemer practice) or arranged around part of the circumference. These rotating converters (of capacity 3–4 tons of pig iron) were confined almost entirely to Swedish practice.

Examples of fixed side blowing converters included the Clapp–Griffiths Converter and an improved version, the Hatton Fixed Converter. Tipping side blown converters included the Robert Converter, the Walrand Legénisel Process vessel and the Tropenas Process vessel. The last three were all designed to produce small quantities of steel possessing great fluidity on account of the high temperature reached during blowing.

The side blown converter developed by Frenchman Alexandre Tropenas was championed in the UK by Edgar Allen & Co. Ltd. of Sheffield for the production of intricate steel castings after the company found that the Robert Converter did not fit the bill. Not to be outdone a near neighbour, Hadfield's Steel Foundry Co. Ltd., installed small converters for similar applications and referred to these as being designed on 'Hadfield's system'.

A further development of the small rotating converter was the Stock Converter, which dispensed with the need for pre-melting the pig iron in a separate cupola prior to refining. The pig was charged cold into a previously heated converter held horizontally, and pressurised oil fuel injected into the region of the blast tuyères. The combustion melted the charge, whereupon the converter was rotated into an upright blowing position, the oil shut off and refining carried out as per a conventional Bessemer blow.

Modern methods

One might be forgiven for thinking that modern bulk oxygen steelmaking practice is greatly removed from the Bessemer process, but this is not really the case.

The VLN process, introduced in the UK by the Steel Company of Wales Ltd., was something of a 'half-way-house' in several respects, effectively bridging the old and new. It was a brave attempt to use Bessemer's pneumatic process to manufacture tonnage dead soft steels suitable for deep drawing and blew an oxygen/steam mixture rather than air to ensure very low nitorgen levels in the refined steels. The acronym VLN was coined from 'very low nitrogen' but the plant was universally known as the 'violin' plant! SCoW's VLN plant commenced production at Abbey Works, Port Talbot, in June 1959 but unfortunately was found to generate large amounts of red oxide fume and was ultimately closed as environmentally unacceptable.

The most widely used tonnage oxygen steelmaking process, the BOS process, employs what is little more than a 'blind' tuyère – less Bessemer vessel where turbulence for rapid refining of the charge is created by the use of supersonic oxygen injected from an overhead water-cooled lance. A variant, the OBM/Q–BOP Process, is able to utilise bottom blowing of oxygen without compromising tuyère refractory life; the simultaneous blowing of hydrocarbon gas causes endothermic 'cracking' around the tuyère regions, where it helps to keep the temperature down.

Even the Swedish Kaldo process can be seen to have some of its roots in the Bessemer process.

Finale?

In many respects the Bessemer process was before its time. Quite simply, the control technology of the 1860s and 1870s was unable to keep pace. This, coupled with a number of inherent disadvantages associated with Bessemer's acid process, meant that the much slower Siemens melting process rapidly gained favour as an alternative means of bulk steelmaking.

Perhaps the area of greatest potential that was never really exploited in mild steel production was the duplex process, i.e. employing the Bessemer converter to rapidly de-siliconise a melt, and then finish the refining by use of a second type of furnace such as open hearth or electric arc.

Eventually this potential of the Bessemer process in duplex working was recognised by the Union Carbide Corporation of America, which developed the use of the converter as part of a production route for making low carbon austenitic stainless steel. However, whereas in earlier duplexing the Bessemer converter was the 'workhorse' at the start of refining, with Union Carbide's process it became the finishing vessel responsible for fine tuning the melt to degrees of compositional control once undreamed of.

Union Carbide's process elegantly solves a 'modern' metallurgical conundrum – how to decarburise a stainless steel melt without losing the potentially more active (and expensive) chromium into the slag. (It is possible to make austenitic stainless steel in the electric arc furnace but chromium recovery from the slag following refining involves time-consuming reversion techniques employing last minute ferro-silicon additions).

Union Carbide's AOD process (an acronym for argon oxygen decarburisation) addresses the problem by reducing the partial pressure of the oxygen injected into the melt by diluting it with argon, an inert carrying gas. This promotes controlled oxidation of carbon in preference to chromium. The AOD technique is employed in conjunction with an electric arc furnace, which is used to provide the basic melt. This is transferred to a Bessemer type vessel, the AOD converter, and decarburised by selective use of oxygen/argon mixtures (oxygen-rich in the initial stages of refining, and argon-rich towards the end).

Technical development of the AOD process took place from the mid-1950s, with the first commercial AOD vessel being commissioned at the Joslin Manufacturing & Supply Company at Fort Wayne in 1968. The UK's largest AOD installation was commissioned in the late 1970s on part of British Steel's stainless steel complex at Shepcote Lane in Sheffield – now part of Avesta Sheffield Ltd.

So, although the orthodox Bessemer converter no longer lights up the night sky of Sheffield's Lower Don Valley, its modern day successor continues to make a massive contribution to the region's stainless steel output – and indeed to Sheffield's continuing metallurgical eminence. At this point one cannot help but hark back to the 1850s, when Sheffield's metallurgists were so hostile to Bessemer and his converter. Perhaps, in some ways, the last laugh is Bessemer's.

Index